GAS CHROMATOGRAPHY IN BIOLOGY AND MEDICINE

GAS CHROMATOGRAPHY IN BIOLOGY AND MEDICINE

A Ciba Foundation Symposium

Edited by
RUTH PORTER

J. & A. CHURCHILL LTD.
104 GLOUCESTER PLACE
LONDON . . . 1969

First published 1969

Containing 67 illustrations

Standard Book Number 7000 1428 4 (cased)
7000 1442 X (limp)

Printed in Great Britain

Contents

* This paper was not read at the meeting but contributed for publication afterwards

Membership

Symposium on Gas Chromatography in Biology and Medicine held 5th and 6th February, 1969.

J. P. Payne (Chairman)	Research Department of Anaesthetics, Royal College of Surgeons of England, London
D. J. Blackmore	RAF Institute of Pathology and Tropical Medicine, Home Office Central Research Establishment, Aldermaston, Berkshire
C. J. W. Brooks	Chemistry Department, The University, Glasgow
H. V. Carter	Perkin-Elmer Ltd, Beaconsfield, Buckinghamshire
F. W. Cervenko	Department of Anaesthesiology, Queen's University, Kingston, Ontario
A. S. Curry	Home Office Central Research Establishment, Aldermaston, Berkshire
S. H. Curry	Department of Pharmacology, The London Hospital Medical College
I. C. Geddes	Department of Anaesthesia, University of Liverpool
G. M. Gray	Lister Institute of Preventive Medicine, University of London
D. W. Hill	Research Department of Anaesthetics, Royal College of Surgeons of England, London
C. Hitchcock	Unilever Research Laboratory, Coleworth House, Sharnbrook, Bedford
J. Janák	Institute of Instrumental Analytical Chemistry, Czechoslovak Academy of Sciences, Brno
G. Levis	Department of Clinical Therapeutics, University of Athens School of Medicine
S. R. Lipsky	Section of Physical Sciences, Yale University School of Medicine, New Haven, Connecticut
H. J. Lowe	Section of Anesthesiology, University of Chicago Medical School, Chicago, Illinois
G. B. Marson	Digital Systems Division, Kent Instruments Ltd, Luton, Bedfordshire

A. J. P. Martin	Abbotsbury, Elstree, Hertfordshire, and Technological University, Eindhoven, Netherlands
F. L. Mitchell	Division of Clinical Chemistry, MRC Clinical Endocrinology Research Unit, Edinburgh*
J. H. Purnell	Department of Chemistry, University College, Singleton Park, Swansea, Wales
M. J. Purves	Department of Physiology, University of Bristol
J. S. Robinson	Department of Anaesthetics, University of Birmingham
R. P. W. Scott	Unilever Research Laboratory, Coleworth House, Sharnbrook, Bedford
J. Sjövall	Kemiska Institutionen, Karolinska Institutet, Stockholm
B. C. H. Warren	Physics Research Division, Chemical Defence Experimental Establishment, Porton Down, Wiltshire

*Present address: Division of Clinical Chemistry, MRC Clinical Research Centre, Harrow, Middlesex.

The Ciba Foundation

The Ciba Foundation was opened in 1949 to promote international cooperation in medical and chemical research. It owes its existence to the generosity of CIBA Ltd, Basle, who, recognizing the obstacles to scientific communication created by war, man's natural secretiveness, disciplinary divisions, academic prejudices, distance, and differences of language, decided to set up a philanthropic institution whose aim would be to overcome such barriers. London was chosen as its site for reasons dictated by the special advantages of English charitable trust law (ensuring the independence of its actions), as well as those of language and geography.

The Foundation's house at 41 Portland Place, London, has become well known to workers in many fields of science. Every year the Foundation organizes six to ten three-day symposia and three to four shorter study groups, all of which are published in book form. Many other scientific meetings are held, organized either by the Foundation or by other groups in need of a meeting place. Accommodation is also provided for scientists visiting London, whether or not they are attending a meeting in the house.

The Foundation's many activities are controlled by a small group of distinguished trustees. Within the general framework of biological science, interpreted in its broadest sense, these activities are well summed up by the motto of the Ciba Foundation: *Consocient Gentes*—let the peoples come together.

CHAIRMAN'S OPENING REMARKS

J. P. Payne

We owe a debt of thanks to the Ciba Foundation because there are few institutions which allow the lack of formality that is achieved in the type of meeting that this Foundation runs. I know of no other place where a small group can meet so informally and yet, at the same time, get through so much work.

I am delighted that all the invited members of the group, except, unfortunately, Professor Garattini, are here, and we are especially honoured that Professor Martin, who tends to avoid meetings about gas chromatography, has been able to join us. Some of you may feel that we have not chosen the members of this group as wisely as you could have done yourselves. You may be missing your friends. But because the total number of scientists who can be invited to the Ciba Foundation's conferences is restricted to twenty-five it is possible that had they been here you might not have been.

The motive behind this symposium was the need for a multidisciplinary approach to the problems of gas chromatography. My predecessor in the Chair of Anaesthetics at the Royal College of Surgeons, Professor R. F. Woolmer, was convinced of the need for such an approach to all medical and scientific problems, and the Research Department of Anaesthetics at the College of Surgeons has followed this multidisciplinary pattern since its foundation in 1957. I was brought up in the same tradition and I have done my best to maintain it. On the grounds that it is no longer possible for any one individual to acquire the detailed knowledge necessary for the continued development even of his own subject, the case for a multidisciplinary approach is overwhelming when different fields of research overlap. Many disciplines are represented at this symposium—anaesthesia, chemistry, engineering, pharmacology, physics and physiology to mention only some—and with this representation we ought to be able to sort out some of the problems which face us.

We want the meeting to be as informal as possible; the papers are only a minor part of the proceedings—the guiding lines for the discussion as it were—and in the next two days we shall need all the time we can find for discussion.

The object of this symposium is to promote the exchange of ideas about gas chromatography and its applications. The responsibility for this exchange now rests with all of us.

HISTORICAL BACKGROUND†

A. J. P. MARTIN

Abbotsbury, Elstree, Hertfordshire,
and Technological University, Eindhoven, Netherlands

I AM going to try to present a logical account of the development of my scientific ideas. When I was a schoolboy I was exceedingly interested in chemistry and read my elder sister's university text books. I do not remember learning any chemistry at all at Bedford School, since I was always well ahead of what I was supposed to be learning in physics and chemistry. I took great interest in distillation and was particularly impressed by distillation columns. By the time I went to Cambridge I had found a number of books describing the chemical engineering side of distillation columns, and noted that at that time (about 1930) industrial research on the preparation of good columns was much in advance of laboratory research. Eighty-plate columns turning out many tons of alcohol or petroleum were available in industry, but distillation systems in laboratories contained, at best, only a few tens of plates.

At Cambridge University I became interested in countercurrent separations and plate theory, and after graduation I started some research in the Nutritional Laboratory at Cambridge. I was interested in vitamins, and decided to look for Vitamin E, unknown at that time. Various members of the laboratory were concerned with the carotenes and in 1933 Dr. A. Winterstein from E. Kühn's laboratory in Heidelberg visited us and demonstrated a chromatogram of a crude carotene solution on a chalk column; the carotene separated appropriately into bands of various colours. I was fascinated to see the relationship between the chromatogram and distillation columns and to realize that the processes involved in the separation of the carotenes and of volatile substances by distillation column were similar; there was relative movement of the two phases and it was their interaction at many points that gave rise to good separations. I continued my work with Vitamin E and

† This presentation was contributed by Professor Martin at the end of the symposium at the special request of the Chairman and all the members of the group. The Editor also wishes to thank Dr. A. T. James, Unilever Research Laboratories, for his help in the preparation of this material for publication.

Dr. T. Moore and I started separating carotene by distribution between two solvents using separating funnels. I was sufficiently mathematically inclined to work out the extent of separation that can be obtained in this way, and was appalled to find how small this was with a single extraction. So I set up chains of separating funnels, moving upper and lower layers countercurrently, but found that even when one has such a small number as, say, six funnels just shaking and separating the layers becomes a full-time job.

I had always been interested in engineering processes and so I started to devise machines to do the countercurrent extractions. The first machine was designed for the first stage in the separation of Vitamin E; vegetable oils were saponified and the soaps extracted with ether. This was a tedious, smelly job. So I put one twenty-litre aspirator bottle on the floor outside the laboratory and another on the flat roof (the laboratory was a single-storey building). I filled the top bottle with soaps and the bottom bottle with ether and joined the bottom of the top bottle to the top of the bottom bottle with half-inch-bore tubing. By this means the ether and soaps changed places over a period of hours (over night, in fact). I found that ten feet of tubing gave about eight theoretical plates and I obtained very efficient extraction in this way. This method was satisfactory for extracting a particular substance from one liquid to another but much more was needed to separate two or more substances of closely similar partition coefficients. It was by no means obvious how one could duplicate the performance of a batch distillation column. Devising a still to evaporate the liquid leaving the column, and continuously dissolving the residue in the other liquid phase was not easy. I can still remember the delight of realizing (while walking home to lunch) that all that was necessary was to inject the substance to be separated into the centre of the column and fix the ratio of the flow rates of the liquids so that they equalled the reciprocal of the partition coefficient. The liquids flowing in at the ends of the column then carried the wanted substance back to the centre of the column and allowed it to escape only very slowly. But other substances, with higher or lower partition coefficients, left more or less rapidly at one end or the other.

It was not worth-while trying to make an apparatus with less than about two hundred theoretical plates, and so I amplified the type of apparatus I had used for the ether extraction of the soaps. Forty-five half-inch tubes, each about five foot long, were stacked vertically in a rack. A pair of narrow tubes ran up between the top of one tube and the bottom of the succeeding tube and this was repeated for the whole series. When a pulse of liquid was pushed through this apparatus the heaviest liquid, which had collected at the bottom of the tube, was forced to the top of the next tube in the series, and a ball valve prevented the liquid from dropping back. Similarly the lightest liquid, which had collected at the top of the tube, was pumped down to the bottom of the next tube (on the reverse stroke) and so on. These successive strokes produced a circulation between the bottom of one tube and the top of the next and made available, in effect, a continuous 200-foot column for separation purposes. The 90 ball valves rattling on their seats made a noise like the sea on shingle!

I used cyclohexane and methanol, or various petroleum ethers with methanol and water mixtures, as solvents. Small drops or saucers, about one millimetre in diameter filled the tubes when the apparatus was working. The machine was completed with diaphragm pumps (the diaphragms were protected from the solvents by mercury) which pumped the liquid through the tubes and from the top and bottom of a single external reservoir at any desired rate; an evaporation system collected the separated substances as they flowed from the ends of the column and returned the solvents to the reservoir. It was a considerable effort to make this machine and when it was finished it was a month before I could summon the courage to try it out.

But apart from some fires and difficulties when tubes broke, the machine did ultimately work. I was able to separate Vitamin E into several obviously different and distinct fractions, for the first time. I have always had difficulty in writing up my results and I never published this work (although it does appear in my Ph.D. thesis) or pursued it further, but it left me with a profound interest in countercurrent problems.

In 1937 my chief, Sir Charles Martin, introduced me to R. L. M. Synge who was working on a scholarship from the International Wool Secretariat at the Biochemical Laboratory at Cambridge. Synge was trying to improve the methods for the analysis of proteins. He had measured the partition coefficients of acetylamino acids between chloroform and water and thought that a separation method could be based on this sort of method. But his technique with separating funnels was not good enough. Sir Charles suggested that my apparatus be used. But chloroform and water were not suitable phases for it, so we designed another completely different machine in which these two compounds could be used. This was made in Cambridge for Synge to my design, but it didn't work, and I took it and Synge to the Wool Industries Research Association, Leeds, where I had moved in 1938. We were eventually successful in using this machine to separate, and measure fairly accurately, the monoamino, monocarboxylic acids in wool. It was a fiendish piece of apparatus, we had to sit by it for a week for one separation; it had 39 theoretical plates and filled the room with chloroform vapour. We used to watch it in 4-hour shifts. We had constantly to adjust small silver baffles to keep the apparatus working properly. One of the effects of 4 hours of chloroform intoxication was that when our partner arrived to take the next shift he was invariably sworn at by the one who had been watching the machine. Another curious effect of the chloroform was that when I went into the fresh air, it smelt peculiar. This was my first experience of the interesting phenomenon of negative smell and may have been partly responsible for my current interest in the physiology of the sensation of smell.

I continued designing new machines that I hoped would be more satisfactory, but although I worked out some dozens of ideas none of them produced a machine that was sufficiently cheap and easy to seem worth making. In 1940 it occurred to me that the crux of the problem was that we were trying to move two liquids in opposite directions simultaneously. Equilibrium had to be established rapidly or the experiment took far too long, but this meant converting the liquids to very fine droplets and if the droplets were too small they would not settle out or move in the required direction within any reasonable period of time. This meant that the machine was bound to be a compromise unless I could either introduce centrifugal force to speed up the movement of the droplets or think of a completely different system. Then I suddenly realized that it was not necessary to move both the liquids; if I just moved one of them the required conditions were fulfilled. I was able to devise a suitable apparatus the very next day, and a modification of this eventually became the partition chromatograph with which we are now familiar. Synge and I took silica gel intended as a drying agent from a balance case, ground it up, sieved it and added water to it. We found that we could add almost its own weight of water to the gel before it became noticeably wet. We put this mixture of silica gel and water into a column, put the acetylamino acids on to the top and poured chloroform down the column. We wondered how we should know where the amino acids were in the column and when to expect them to emerge at the bottom of the tube. By the end of the first day there was no sign of them. To find out what was happening in the column we added methyl orange to the liquid on the silica gel and thus were able to see the acetylamino acids passing down the column as a red band. One foot of tubing in this apparatus could do substantially better separations than all the machinery we had constructed until then.

Normal chloroform contains about 1 per cent of ethyl alcohol as a stabilizer. The first experiment we did, with chloroform straight out of a bottle in the laboratory, gave the results we expected, and we separated acetylalanine and acetylleucine. We next used carefully distilled chloroform and were surprised to find that the amino acids did not move from the top of the column. The reason for this, of course, was increased absorption due to the absence of ethyl alcohol. When we added alcohol to the chloroform our bands could move down the column again. But it was difficult to produce a satisfactory colour change; we needed large amounts of acids with our original silical gel-methyl orange system. So we experimented with different ways of making precipitated silica, and eventually developed a process of stirring hydrochloric acid into sodium silicate. This process reliably produced material that behaved as we wished in the columns. But this work was more magic than science; we never understood in detail what we were doing. Later we changed the indicator; at one time we used pelargonidin which we extracted ourselves from various flowers.

In spite of all our efforts we could separate only the monoamino, monocarboxylic acids. We could not make the system work for the dicarboxylic or basic amino acids. So we looked for materials other than silica to hold the water, and our first choice was paper. I had seen paper chromatograms of dyes and was familiar with the uptake of water by cellulose, so paper was an obvious choice. Dr. A. H. Gordon, who was now working with us at Leeds, looked through *Beilstein's Handbuch der Organischen Chemie* to find a colour reaction that would reveal our amino acids on the paper; he found ninhydrin which proved admirable for our purpose. Our first paper chromatograms were circles of paper in a Petri dish containing water and water-saturated butanol fed by capillarity to the centre by a tail on which a drop of amino acid solution had been placed. When the butanol reached the edge, the paper was dried and sprayed with ninhydrin in dry butanol. Later, we used strips of paper in test-tubes and more suitable containers—boxes in which the air was kept saturated with water—with troughs containing the mobile solvent into which the tops of the strips could dip. Several boxes were needed since it was characteristic of the method that though it was not particularly quick, very little work was needed to run many strips simultaneously. An important step was running the chromatogram in two dimensions. The first solvent spread the amino acids in a line near one end of the paper from a spot near the corner; then, after drying, we turned the paper through a right angle and spread the line of spots into a two dimensional pattern by using a different solvent. (See Consden, Gordon and Martin, 1944.)

Our next problem was to deal with the curious fact that in some, but not other, solvents the purple amino acid spots had a pink "beard" underneath them; and as they ran further down the paper the purple colour showed less and the pink more. The purple-coloured spots of leucine and phenylalanine, for example, had almost vanished before they got to the bottom of the paper, leaving only a faint pink blob. This unsatisfactory colour change was particularly marked with papers on which the chromatogram had been run in two directions. We could not understand the reason for this. We first tried running in one direction in phenol and in the other in collidine, which can distinguish between the acidic, basic and neutral compounds. (The first two-directional separation that I did was with electrophoresis in one direction in an acetate buffer and paper chromatography in the other. The tract for the electrophoresis was isolated from the rest of the column by saturating the paper with paraffin wax on either side. But chromatography turned out to be more satisfactory than electrophoresis at that time, so I did not work with electrophoresis again until 1944.) Eventually we found the cause of these pink beards. Phenol was used as one of the solvents and when the separation was run in an atmosphere of ammonia so as to increase the pH, the paper became covered with black spots. We identified the cause of these spots as copper from the fans used for drying the papers in the laboratory; these fans had a badly sparking commutator which filled the room with copper. The large amounts of copper in the Leeds atmosphere also contributed to the copper on our papers. We finally discovered that the pink beards were caused by a copper complex of the amino acid that formed on the paper. The black colouration, due to the catalytic oxidation of phenol by copper in the presence of ammonia, indicated that copper was present. The formation of these copper complexes could be suppressed by including a complexing agent for copper—cyanide for example—in the coal gas we put into the atmosphere in the box. This technique was gradually developed into the paper chromatographic systems that are used today for the separation of amino acids.

We were pleased to find that our system worked equally well for peptides and surprised to find that it could separate almost every other group of compounds it was tried on—carbohydrates (Partridge, 1946) and flower petal anthocyanin (Bate-Smith, 1948*a*, *b*) are examples. F. H. Pollard (at Bristol) worked with metals and found that they also could be separated by paper chromatography. Synge and I were busy with the amino acids and peptides, but we were visited by many scientists who were interested in our technique. They came and looked at the paper chromatograms and then went away and used paper chromatography for their own separations. Paper chromatography was amazingly applicable to the separation of widely different groups of chemicals.

This work was eventually published in 1941 (Martin and Synge, 1941). In this paper we noted that the mobile phase could just as well be a gas as a liquid. We also predicted that, if the stationary phase were a liquid, very refined separations of various kinds of compounds would be possible. Although this paper was widely read by chemists in the petroleum industry, no one thought this prediction worth testing experimentally until, nine years later, in 1950, Dr. A. T. James and I started to work on gas liquid chromatography (James and Martin, 1951, 1952, 1956). Synge and I, in our first paper on partition chromatography (Martin and Synge, 1941), had evolved a theory relating the speed of the zones to the partition coefficient. Further, by introducing the concept of the theoretical plate for chromatograms, a prediction could be made about the shape of the zones and their rate of broadening. Later, after work with peptide paper chromatograms, I found it possible, by assuming that the free energy of transfer of a compound from one phase to another was an additive function of the free energies of individual atoms or groups of atoms, to forecast with reasonable accuracy the partition coefficient and chromatographic behaviour of peptides and many other substances.

In 1948 I moved for a short time to the Lister Institute of Preventive Medicine in London and then to the National Institute for Medical Research, Mill Hill, where Tony James later joined me (he had previously been working with Synge). James and I tried to separate our materials using crystallization on a column—what is now known as zone-refining. This project looked hopeless for a few months, and James became more and more discouraged; we could do much better with a couple of beakers than with all the complicated apparatus we had constructed. So (to improve James' morale) I suggested that we study gas chromatography; I was sure this would work. Professor J. Popjak had asked me for a more refined method than paper chromatography for separating fatty acids and I thought that gas chromatography might be able to do this. So we spent our first week waiting for the bands to come out of a gas chromatograph: in fact, they had all come out in the first few seconds. We used quarter-inch-bore glass tubing, about 15 inches long, packed with Celite (which had been found to be the most convenient material to use with liquid–liquid columns). We passed nitrogen in at one end of the column, the other end of which was provided with a capillary that dipped into a test-tube containing indicator solution. A small conical flask, instead of a burette, held the titrant. The flask had a doubly bored stopper, one hole carrying a tube that passed from the bottom of the flask to a jet just above the level of the liquid in the test-tube, while the other hole had a piece of valve rubber attached that could be milked between finger and thumb to express a drop of titrant from the jet.

James sat with a stop-watch and a piece of graph paper and timed and plotted the drops while I watched the test-tube and put in a drop of titrant whenever the colour of the indicator changed. Plotting the number of the drops against time yielded a series of steps. The height of the steps denoted the quantity of acid emerging, and their position on the time axis showed the retention time. We first separated the methylamines, since they would run at room temperature. Later, using a steam jacket for the column, we separated the first members of the fatty acids series. Initially we used a fatty, oily material, but this gave very distorted bands. I had enough experience with chromatography by this time to realize that non-linear absorption creates tailing. But on these plots we had the reverse of tailing—a long front and a sharp tail—which we eventually realized was due to the association of the fatty acid to dimers; in other words, dimerization was a considerable problem to us for six months or more. By adding a soluble acid (such as stearic acid) in excess to the stationary-phase liquid, we were able to sort out this difficulty and obtain reasonably shaped bands. This technique worked very well for fats and oils and equally well for amines. We obtained our first useful results six weeks after starting the experiment. This was the beginning of gas chromatography for us. The details still had to be worked out, but it is really astonishing how closely similar this first column was to many columns still in use today. We wanted to illustrate the technique by using it to separate some natural mixtures, so we tried to identify the amine responsible for the fishy smell of stinking goose-foot (*Chenopodium vulvarium*). We found trimethylamine in this plant and were able to separate the three methylamines and ammonia quite readily from it.

We next made an automatic titrating machine. This was a rather Heath-Robinson arrangement. The titration was recorded by using the eye to detect colour changes in the indicator and the whole thing was operated by manual drive. This was a most demanding machine; if one looked away for a second a kink in the curve appeared! We next incorporated a motor with a photocell, to drive the machine automatically. In 1952 Dr. R. P. W. Scott, from the Research Laboratories of Imperial Chemical Industries Ltd, consulted James and me about the separation of hydrocarbons. We suggested using a thermal conductivity cell because S. Claesson (1952) had already described displacement of gas chromatography on charcoal, using a conductivity cell. In 1953, Dr. N. H. Ray told us about his results with the thermal conductivity cell (Ray, 1954), but his method was not completely satisfactory for us. I tried to devise a better method and developed the gas density balance (James and Martin, 1956), a good detector in its day.

That is really my only contribution to gas chromatography; since 1955 I have worked in other areas although I am now again working on electrophoresis. Tony James and I studied enough different systems to show the kinds of relationships that can occur in liquid chromatography, that is, we found that there is a log relationship between retention volumes and the numbers of carbon atoms in a molecule and we probably laid the theoretical foundations for gas liquid chromatography reasonably well at that time. Partition chromatography has developed much further than Synge and I originally expected, perhaps most surprisingly in connexion with the quantity of material needed for analysis. Accepted methods of amino acid analysis before we began our work required half a kilogramme of protein and about six months' work for a monoamino, monocarboxylic analysis. The silica partition columns require a few milligrammes and paper chromatographs only a few microgrammes of protein. Now gas chromatographs with modern detectors can work with nanogramme quantities; these methods are indeed the most sensitive that are available so far for the analysis of many substances. Thus, within a thirty-year period, the quantity of sample needed has been reduced by a factor of 10^{12}. I am hopeful that methods I hope to work on, either myself or through others, will reduce this quantity by a further 10^3–10^6 without loss of accuracy.

REFERENCES

Bate-Smith, E. C. (1948*a*). *Nature, Lond.*, **161**, 153.
Bate-Smith, E. C. (1948*b*). *Nature, Lond.*, **161**, 835–836.
Claesson, S. (1952). *Ark. Kemi. Miner. Geol.*, **23A**, 1.
Consden, R., Gordon, A. H., and Martin, A. J. P. (1944). *Biochem. J.*, **38**, 224.
James, A. T., and Martin, A. J. P. (1951). *Biochem. J.*, **48**, PVII (abstract).
James, A. T., and Martin, A. J. P. (1952). *Biochem. J.*, **50**, 679–690.
James, A. T., and Martin, A. J. P. (1956). *Biochem. J.*, **63**, 138–143.
Martin, A. J. P., and Synge, R. L. M. (1941). *Biochem. J.*, **35**, 1358.
Partridge, S. M. (1946). *Nature, Lond.*, **158**, 270–271.
Ray, N. H. (1954). *J. appl. Chem.*, **4**, 21, 82.

GAS CHROMATOGRAPHY: THE ANATOMY OF A SCIENTIFIC REVOLUTION

S. R. LIPSKY

Section of Physical Sciences, Yale University School of Medicine, New Haven, Connecticut

IN the preface to the first edition of one of the earliest books on chromatography, *Principles and Practice of Chromatography* (Zechmeister and Cholnoky, 1943) there is written: "Every scientific advance is an advance in method." It is also stated in this preface that "the invention of a new specialized laboratory procedure brings about rapid conquests in new fields of science and technology, finally it exhausts itself and is replaced by a still more practical method. The method of chromatographic adsorption invented by the talented Russian botanist, Professor M. Tswett makes possible spatial separation of components of a mixture. It is just now at the beginning of a bullish development: it offers a simple experimental procedure to the investigator especially in the fields of both pure and applied organic chemistry, of biochemistry and of physiology."

Undoubtedly, today, much of the content of this statement would be appropriate to the most powerful technique ever devised—gas chromatography. One could dwell at great length on all the developments that have contributed to make this so. Moreover it would be difficult to think of any scientific discipline which has not been affected by this discovery. The unique feature of the method that made repeated breakthroughs possible was the fact that the sample remained in the gas phase and the resistance to mass transfer in the mobile phase was relatively small. Since the original contributions of Professor Martin (Martin and Porter, 1951; James and Martin, 1951, 1952) we have learned much about the transfer processes that occur in gases, and much of our present theory and column and detector technology have evolved with remarkable speed because of this.

With the maturing of the field of gas chromatography, and the inevitable diminution of the number of solid scientific advances in this sphere with the passage of time, a few theoreticians and experimentalists continued to look for new horizons. Naturally they turned their energies to an area that had been with us since the beginning of this century—liquid chromatography. The problems were easily defined, but their solutions have proved more elusive. In essence the main problem was this: Could those molecules or even their derivatives which are not readily analysed by gas chromatography (compounds with high molecular weights or the highly polar materials that do not possess sufficient vapour pressure to be volatilized without thermal rearrangement or degradation) be separated in the *liquid phase* by techniques possessing the same versatility, speed, resolution and sensitivity that we have come to expect in *gas-phase* analysis? If this were so, it would bring about a magnificent era characterized by challenges of unparalleled magnitude. And it would profoundly influence a large number of different scientific disciplines. For example, in biochemical research, the consequences would be revolutionary if we could rapidly separate and detect picogramme quantities of all the different forms of insulin, each having a slightly different amino acid sequence and each easily synthesized by the solid-phase method of Merrifield (1963). This would be a daring step forward for the scientifically oriented physician. He would at last be in a position to precisely correlate the relationship between chemical structure and physiological function of the small active peptides and proteins.

Professor Howard Purnell wants to take steps to eliminate the column from gas chromatography and perhaps introduce a new form of spectrometry. I would like, via the concept of a different sort of column, to travel in a rather similar direction. I hope that these two routes may lead to the same desired goal—the easy and rapid separation and detection of picogramme quantities of all chemicals—without meeting too many insoluble problems on the way.

Recently, scientists were awed by the synthesis of ribonuclease, one of the most important enzymes known to man, by the Rockefeller (Gutte and Merrifield, 1969) and Merck (Denkewalter *et al.*, 1969; Strachan *et al.*, 1969) groups in the US. This enzyme, made up of 124 amino acids, controls the composition and distribution of vital organic compounds in cells. Now that the barriers to synthesis have finally been overcome, scientists will eventually be able to make an infinite variety of ribonucleases containing a full range of differences (from subtle to profound) in amino acid sequences. Each ribonuclease thus produced will need purification and precise structural identification before being assessed for physiological function in a variety of systems.

The solution of this problem is one of the most important challenges of our time. Those of us with long experience in gas chromatography, where the separation of structurally similar compounds in microgramme to femtogramme quantities is accomplished in minutes, cannot doubt that knowledge gained in this field will, in time, be extrapolated to these new and vitally important areas. When we remember the surprise of the sceptics when positional, configurational and optical isomers were conveniently analysed by gas chromatography, we can surmise that similar separations by liquid chromatography will be accomplished in the next five to ten years. Thus I am sure that techniques will soon become available whereby the separation of the individual components of mixtures of complex macromolecules (mol. wt. 2000–10 000) will be achieved on the basis of very slight differences in their chemical structure.

At the present time, developments in the field are proceeding satisfactorily and the various obstacles are being gradually surmounted. Thus, for example, due mainly to the efforts of Giddings (1966), Horvath, Preiss and Lipsky (1967), Horvath and Lipsky (1969*a*, *b*), Huber (1969), Knox (1966), Pretorius and Smuts (1966), R. P. W. Scott and his group (Scott, Blackburn and Wilkins, 1967) and my own group, one of the first problems to be resolved was the development of methods for overcoming the extraordinary resistance to mass transfer in the liquid phase when compared to that in the gaseous state. One of the ways in which this difficulty is being circumvented, at least to some extent, in our laboratories by Horvath and myself (Horvath, Preiss and Lipsky, 1967; Horvath and Lipsky, 1969*a*, *b*) is by the use of narrow-bore columns (diameter: 5·0–7·5 mm., length: 1·0–3·0 m.) packed with glass beads that are coated with a pellicle irreversibly bound to the sphere in order to withstand the relatively high pressures (3·04–18·24 MN/m.^{2} = 30–500 atm.) at which we operate the system. The pellicle (the stationary phase) may consist of various prescribed thicknesses of adsorbent, cation or anion exchange resin, or any other suitable stationary phase that can be appropriately fixed to the glass surface. In these circumstances we can readily and rapidly utilize a wide range of buffers at different pH values and temperatures with or without gradient elution in order to achieve rapid analysis. Thus, for the first time, we have been able to separate quantitatively nanomolecular amounts of 18 or more individual components in a mixture of the mono-, di- and triphosphate ribonucleosides in 90 minutes. Previous methods required up to 30 hours to accomplish this and did not even approach the degree of resolution achieved here. Moreover, a similar system was used to isolate and accurately detect the bases of nucleic acids at the picomolecular level in less than 5 minutes (Horvath and Lipsky, 1969*b*). When applied to appropriate extracts of mouse brain and liver a phenomenon familiar to gas chromatographers working with biological systems was observed: about 15–20 peaks representing ultraviolet (u.v.) adsorbing materials were noted, whereas only five or six components emerged when more conventional methods were used. If we had had a more universal sensing system, additional eluted bands might have been detected.

This brings us to a fascinating question: What was the precise chemical structure of each of the components that were isolated? These were present in sub-microgramme quantities. They exist in the liquid phase. Obviously they cannot be readily identified by the uniquely sensitive techniques now accepted as commonplace for detecting organic compounds in the gas phase, namely combined gas chromatography and mass spectrometry, gas chromatography-infra-red (i.r.) methods and so on.

Although the nuclear magnetic resonance (n.m.r.) spectrometer is a most powerful tool for providing us with definitive structural information, it requires samples in the 50–500 microgramme range—an often appalling requirement for the biochemist. Nonetheless, here too there is reason for optimism. If Fourier transform analysis (Ernst and Anderson, 1966) is utilized, it is logical to assume that we can decrease the sample size by a factor of at least 10 to 20 and analysis time by a factor of 100. But if we are aiming at eventually analysing the nucleic acids or proteins of a single cell or cluster of cells, even n.m.r. is not sensitive enough.

At this point it is useful to re-examine the entire concept of the detection of nanomole quantities of multicomponent mixtures in order to see if we need an entirely new form of spectrometry (in the broadest sense) to achieve our goals. It is probably too early to decide this at the moment; some form of chromatography linked with a unique identification technique may be what is needed. Certainly separation of complex mixtures into individual chemical entities prior to structural analysis will provide less difficulties in the interpretation of data from such structural information systems. Even today, with ingenuity, hard work and luck we may be able to utilize some recent developments in other scientific disciplies. Electron emission spectroscopy, although in its early stages, may have enormous potential in this area. It is certainly encouraging to learn that sample sizes (solids) as small as 10^{-8} gramme can be utilized by this technique. Mass spectrometry, also, may again serve us well here.

Dole, Mack and Hines (1968) have reported that dilute solutions of polystyrene (mol. wt.: 44 000 and 400 000) have been detected by a Faraday cage arrangement after being sprayed into a high vacuum system by a unique sample-introduction system. A nozzle and skimmer were used and the "droplets" were negatively charged before their entry into the field. Under these conditions aggregation was prevented since the charged macro-ions repelled one another. In essence, each droplet contained one discrete macromolecule. This development could have profound ramifications if we could overcome some of our present difficulties, such as the need for the accumulation of sufficient quantities of macro-ions to satisfy our detection systems and also the need for a way of inducing meaningful fragmentation of macromolecules so as to provide definitive structural information. It is not now unusual to obtain the amino acid sequences of a few microgrammes of certain small peptides in derivative form which can be volatilized in the ion source of the mass spectrometer. In the future we should be able to take all the different insulins that have already been separated from one another, place a few picogrammes of each into a unique spectrometer and obtain the amino acid sequence within seconds.

In conclusion, I would like to pay tribute to the man whose genius was most responsible for it all, Professor A. J. P. Martin. He has twice made outstanding contributions to this field, in his discovery of partition chromatography and in his pioneering work on gas chromatography. He has thus altered for the better the lives of many of us. We, his scientific colleagues, thank him for allowing us to share with him this wonderful adventure.

REFERENCES

DENKEWALTER, R. G., VEBER, D. F., HOLLY, F. W., and HIRSCHMANN, R. (1969). *J. Am. chem. Soc.*, **91,** 502–503.

DOLE, M., MACK, L. L., and HINES, R. L. (1968). *J. chem. Phys.*, **49,** 2240–2249.

ERNST, R. R., and ANDERSON, W. H. (1966). *J. scient. Instrum.*, **37,** 93–102.

GIDDINGS, J. C. (1966). *Dynamics of Chromatography*, New York: Dekker.

GUTTE, B., and MERRIFIELD, R. B. (1969). *J. Am. chem. Soc.*, **91,** 501–502.

HORVATH, C. G., and LIPSKY, S. R. (1969*a*). In *Advances in Chromatography*, pp. 268–275, ed. Zlatkis, A. Evanston, Ill.: Preston Technical Abstract Co.

HORVATH, C. G., and LIPSKY, S. R. (1969*b*). *Analyt. Chem.*, in press.

HORVATH, C. G., PREISS, B. A., and LIPSKY, S. R. (1967). *Analyt. Chem.*, **39,** 1422–1428.

HUBER, J. F. K. (1969). In *Advances in Chromatography*, pp. 283–288, ed. Zlatkis, A. Evanston, Ill.: Preston Technical Abstracts Co.

JAMES, A. T., and MARTIN, A. J. P. (1951). *Biochem. J.*, **48,** *PVII.*

JAMES, A. T., and MARTIN, A. J. P. (1952). *Biochem. J.*, **50,** 679–690.

KNOX, J. H. (1966). *Analyt. Chem.*, **38,** 253.

MARTIN, A. J. P., and PORTER, R. R. (1951). *Biochem. J.*, **49,** 215–218.

MERRIFIELD, R. B. (1963). *J. Am. chem. Soc.*, **85,** 2149.

PRETORIUS, V., and SMUTS, T. W. (1966). *Analyt. Chem.*, **38,** 274.

SCOTT, R. P. W., BLACKBURN, D. W. J., and WILKINS, T. (1967). In *Advances in Gas Chromatography*, p. 160, ed. Zlatkis, A. Evanston, Ill.: Preston Technical Abstracts Co.

STRACHAN, R. G., and twelve co-workers (1969). *J. Am. chem. Soc.*, **91,** 503–505.

ZECHMEISTER, L., and CHOLNOKY, L. (1943). In *Principles and Practice of Chromatography*, p. 1. New York: Wiley-Interscience.

DISCUSSION

Payne: Professor Lipsky has outlined the history of the development of gas chromatography; he has posed some problems and offered some solutions and some non-solutions to these problems. Would anyone like to agree or disagree with any of his comments?

Martin: I do not feel hopeful that spectrometric methods will ever be refined enough to elicit much information about large molecules. In order to split such molecules into small enough parts for these to be recognizable one would have to start with a relatively enormous quantity of material because thousands of different types of fragment would have to be identified in order to reconstruct a protein. Spectrometry, therefore, is unlikely to produce the amount of information we need. We shall have to wait for the development of more powerful methods for the identification of the components of macromolecules. Mechanical techniques may eventually be capable of providing this type of information. Although, during the past five years, I have been unsuccessful in my attempts to evolve such mechanical methods, I still believe this to be the best approach. It should eventually be possible to make micromanipulators that can handle single molecules. In my view this is the only way in which it will be possible to identify very small quantities of very large molecules.

Lipsky: We have two problems: separation and identification. We must separate closely related macromolecules from one another before our identification systems can provide the data we need. When we have broken down the large molecules into smaller parts we still have the problem of determining the precise carbon–hydrogen–oxygen or other hetero-atom arrangement, within each moiety, without further degrading it.

Martin: Obviously, if one could handle individual molecules, and had enough time, one would not need the separation step at all; but in investigating a particular field the most efficient separation step available should be used so as to reduce the number of molecules one needs to examine in order to find the interesting one. It would be frustrating indeed to spend a long time taking a cell to pieces if all one could find in it were glucose and water. There should eventually be no difficulty in investigating very complicated large molecules by mechanical means, but obviously this work is still far in the future.

Lipsky: The organic synthetic chemists have recently taken a major step in this area with Merrifield's technique (Gutte and Merrifield, 1969) and they are now looking for more refined methods for the rapid identification of new synthetic products. Slight alterations in structure are now possible, which means that scientists in the pharmaceutical industry and research workers in medicine will soon be able to provide previously unavailable information by testing the physiological functions of discrete, well-defined molecules. Pressure in future will be on those concerned with the development of techniques for the efficient separation and precise identification of organic compounds of high molecular weight.

Scott: I am not convinced that large molecules cannot be broken down in a mass spectrometer in amounts sufficient for identification. With our present MS 9-MS 12 (Associated Electrical Industries [AEI] models) system, which employs molecular multipliers, a few hundred to a thousand fragmented molecules of material of high molecular weight can provide adequate spectra. We use solid injection probes for materials with large molecular weights (for example polypeptides) and we can provide fragmentation patterns that allow amino acid sequences to be determined.

Martin: I agree that this technique might be useful if some method existed for breaking the molecule at any chosen site, but I see no possibility of this at the moment.

Lipsky: Some of the small peptides (containing from 10 to 12 amino acids) certainly fragment in the mass spectrometer in discrete patterns. Of course there are a few alternative patterns of cleavage, but these could be easily picked up by computer techniques. Only about 20 amino acids have to be considered, and by looking at the particular masses in certain areas one could work out which amino acid was present and thus readily deduce the entire amino acid sequence. The technology of mass spectrometry might develop in the following way: if we could introduce new sampling systems to convert macromolecules to, say, the gaseous state, and thus were able to cause their fragmentation in a meaningful manner, industrial chemists would be stimulated to build mass spectrometers with extended mass ranges. The mass range of instruments available today encompasses molecular weights of up to approximately 6000. We can hope that mass spectrometers handling molecules with molecular weights of up to 50 000 will be developed in the near future.

Martin: But one rapidly reaches numbers that make these techniques impractical as, for example, with gas chromatography of the hydrocarbons. Separation of the isomers stops at around twelve carbon atoms but the total number of permutations from this number of isomers is astronomical. We cannot hope for success with mass spectrometry because this technique can never provide the final structure of macromolecules since the mass spectrometer cannot handle a single molecule.

Purnell: Another problem is that one often has to deal with molecules of such similar structure that the fragmentation patterns look alike. We have recently tried to use mass spectrometry for the identification of carbosilanes but this investigation told us nothing because all the fragmentation patterns were virtually identical. We ultimately identified our compounds using gas chromatographic techniques. The basic problem with the mass spectrometer as an identifying instrument is that all it gives us is a value for a mass. Although this is a unique piece of information nothing more can be got out of it than this single number whereas other techniques (gas chromatography and nuclear magnetic resonance [n.m.r.]) yield several pieces of information from which a great deal can be deduced in the long run. This is why I have never been as optimistic as many physical chemists about the use of mass spectrometers for identification purposes.

Scott: I am not familiar with the spectra of the carbosilanes but I am familiar with spectra for terpenes; these show very similar fragmentation patterns for different molecular structures. But there are subtle differences in peak heights, although fragments of the same masses are obtained, and the relative peak heights can be used for purposes of identification. We are identifying terpenes by comparison of spectra, using computer programmes that utilize both peak masses and peak heights. The fragments produced from proteins, of course, are not completely unknown moieties; they must consist of amino acid residues and, further, only a limited number of amino acids are likely to occur. Although I agree that there are problems in using mass spectrometry in a completely unknown field, these—particularly with amino acids—will be significantly reduced as more background knowledge of the spectra becomes available. For the moment I would rather use mass spectrometry than other techniques to elucidate molecular structure, extrapolating from known mass spectra when necessary.

Purnell: There is a large gap in the technology of identification of macromolecules and even of simpler molecules. For the past six years I have maintained, and still maintain, that what we most need is a machine that prints out the name of the substance as it emerges from a column.

Scott: Many instrument companies interested in the gas chromatography–mass spectrometry–computer combination are trying to develop a machine that can do this.

Purnell: I have not yet seen such an instrument.

A typical problem we encounter in gas kinetics relates to pyrolyses. For example, with propylene one gets 26 products, and many of these are found only in sub-nanogramme amounts in conventional experiments. The most sensitive mass spectrometer currently available is probably the MS 9 (AEI model), but even this is insufficiently sensitive to register these minute amounts. I am not decrying what mass spectrometry can do but I do not think it will ever be able to do much more than it is doing now. It cannot offer us the means of taking the giant step we need to fill this technological gap.

Scott: The normal ionizing system used in the mass spectrometer has an ionization efficiency of about 0·00015 per cent. Considerable scope exists for improving sensitivity by increasing the ionizing efficiency. In 1964 (unpublished material) I started (but never finished) looking at the possibility of using metastable atoms to produce ions in the mass spectrometer. Ions can indeed be produced in this way but I have never succeeded in diverting them into the focusing and accelerating system of the mass spectrometer. With our present ionizing system we have a wide range of sensitivity but, due to the low ionizing efficiency, we still throw away most of our sample even after we have placed it in the mass spectrometer.

Purnell: I accept that a major problem is the wastage of material that is thrown down the pump.

Warren: Do *cis*-isomers and *trans*-isomers have a characteristic fragmentation pattern and can they be differentiated from each other?

Scott: Sometimes they can; it depends on the type of fragmentation. Problems with the identification of such materials (as with the terpenes) is that we do not have enough reference data. With i.r. spectrometry there are about 15 000 spectra reported, but with mass spectrometry probably only 3000–4000 reference spectra are available and these are largely of hydrocarbons.

Sjövall: The difference between isomers may be quite striking. The eight stereo-isomers of 3(α,β)-5(α,β), 17(α,β)-pregnan-20-one all give different mass spectra; they have the same fragmentation pattern but the relative intensities of the peaks are reproducibly different.

Lipsky: N.m.r.–mass spectrometry is probably the most powerful combination yet available. But even with this the problem of the difference in sample sizes—the discrepancy between nanogramme quantities in the mass spectrometer and the 100-microgramme amounts needed for n.m.r.—exists. An entirely new sort of identification technique is desperately needed. Such a technique will be developed in the next five or ten years because of recent progress in the synthesis of macromolecules, many of which have significant biological activity. The main hold-up is in the development of sophisticated and rapid techniques for purification, separation and identification.

Sjövall: Speaking as a clinical chemist who does not work with macromolecules, the most important development would be to devise an ion source giving a much higher yield of ions than any source currently available.

Lipsky: The mass spectrometry industry inhibits progress. Mass spectrometers are costly instruments; one is very glad to have one in the laboratory and delighted when it works efficiently most of the time. Beacuse one does not dare to fiddle with such expensive and unpredictable instruments one is dependent on the industry itself for new developments. Unfortunately, markets are established and profitable, and industry does not have the ambition that we might have to make changes.

Scott: I agree. One promising line of research would be to study the production of ions by metastable argon. Metastable argon atoms have an energy of 11·2 eV and in a suitable ionization chamber could be made to produce parent molecular ions only. In these circumstances the whole mixture could be fed into a mass spectrometer and in the resulting spectra each peak would represent the parent ion of an individual component. Then we could really throw away the column. The problem is not ionization but focusing the ion into the right part of the mass spectrometer. To develop this research would require the use of a mass spectrometer for months at a time, which would only be possible in a commerical laboratory.

REFERENCES

GUTTE, B., and MERRIFIELD, R. B. (1969). *J. Am. chem. Soc.*, **91,** 501–502.

THE MAIN AIM OF CHROMATOGRAPHY; ELIMINATION OF THE COLUMN

J. H. PURNELL

Department of Chemistry, University College, Singleton Park, Swansea, Wales

THE chromatographic column is a remarkably simple device of elegant function and enormous analytical power. In contrast, the chromatograph is an ungainly instrument, the numerous parts of which are complex in operation and are brought together in unsubtle fashion. Paradoxically, however, it is the column which presents most problems to the practising analyst. At its simplest, the explanation lies in the "physical" nature of the hardware and the "chemical" basis of column behaviour. The absence of adequate, straightforward theories of solution and adsorption allows little chance of prediction of column performance even by academic experts and to many, in consequence, the choice of column and its operating conditions suggest black magic. In fact, the literature contains an enormous fund of useful and practical information on this subject and it is this which I have been invited to review. In the space available I can do no more than indicate the present position and attempt to outline a philosophy.

The effectiveness of a chromatographic separation is essentially determined by the column, the performance of which is described in terms of (*a*) its *efficiency*, that is, its capacity to produce elution bands of adequate narrowness, and (*b*) its *selectivity*, which is a measure of the extent to which bands are separated in time.

The efficiency is quantitatively defined by the theoretical plate height, H, which in turn is the column length, L, divided by the theoretical plate equivalent of the column, N. The latter is measured from the chromatogram and computed via, for example, the equation

$$N = 5 \cdot 5 \left(\frac{V_R}{d} \right)^2 \qquad (1)$$

where V_R is the volume flow between injection of a substance and appearance of its peak maximum and d is the peak width at half height in the same units as V_R. These may be volume, time, chart-units or any other convenient units.

The selectivity is best defined in terms of the relative "true "retention of a pair of peaks by the expression

$$\alpha = V'_{R2}/V'_{R1} \tag{2}$$

where the number subscript designates the order of elution and the prime indicates retention measurement, not from injection but from appearance of an unsorbed substance, usually air. α, of course, has meaning only for a given pair of substances but it can readily be tabulated for a complex mixture relative to a common substance assigned the value unity. It is worth noting here that N, also, varies from substance to substance and so we see that a considerable element of approximation may enter into any chromatographic computation.

Studies of column efficiency

Column efficiency has been the subject of very intense study for a dozen years or so. The theoretical plate height is found to be described by an equation of the form,

$$H = A + B/u + Cu \tag{3}$$

Although about a dozen variants of this equation have been proposed from time to time, the simple form above is found to be quite adequate for practical purposes.

The A term represents an eddy (dispersion) term about which there is much controversy. In capillary columns it is zero and some theories demand that it be zero for packed columns also. But this is not important since A cannot exceed one particle diameter in magnitude.

The term B/u is the determining quantity at low carrier gas velocity, u, rising towards infinity as $u \rightarrow 0$. B is a function of the interdiffusion coefficient of the sample vapour in the carrier gas, D_g, and has a maximum value of $2D_g$. It can obviously be diminished by use of dense carrier gases or high average pressures. However, the whole picture is somewhat more complex than this simple analysis would indicate.

The term Cu is of greatest importance at high velocity and C represents a summation of resistance to mass transfer between and across phases. In the simplest situation it can be regarded as the sum of two terms, C_g and C_l, which represent inverse first-order rate constants for equilibration of material radially across the gas phase, and through any liquid present, respectively. C_g is inversely proportional to D_g and so reduction of B leads to increase of C_g.

A plot of experimentally measured H versus u has the hyperbolic form illustrated in Fig. 1. As would be expected from the form of equation (3) the curve has a minimum which, if for simplicity we take $A = 0$, is defined by:

$$u_{min.} = (B/C)^{\frac{1}{2}}, \qquad H_{min.} = 2(BC)^{\frac{1}{2}}$$

It is common practice to determine the location of this minimum for a column and to operate there. The minimum varies somewhat from substance to substance but a good compromise is usually possible.

The above equations allow some estimate of the attainable least value of H_{min}. From the definition of B we can hardly expect it to be much smaller than

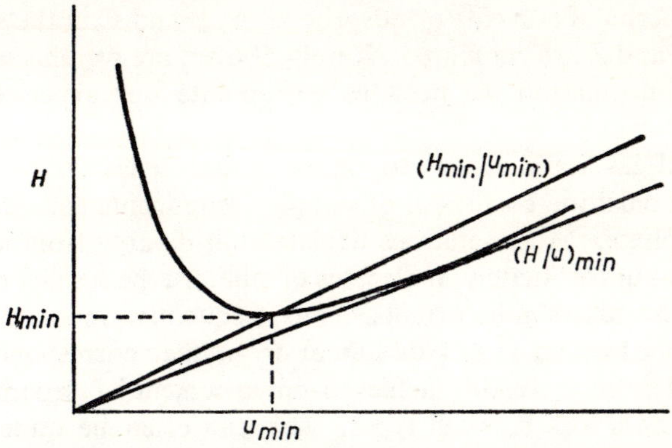

FIG. 1. Typical graph of theoretical plate height, H, plotted against carrier gas velocity, u. Note that the line $(H/u)_{min.}$ has a smaller slope than does the line $(H_{min.}/u_{min.})$. Since analysis time is proportional to H/u, it may be better to work along $(H/u)_{min}$.

about 10^{-1} cm.^{2} sec.$^{-1}$ while an excellent value of C for a packed column of 140 mesh support would be 5×10^{-4} sec. Thus, the least value of $H_{min.}$ practically realizable is about $1{\cdot}4 \times 10^{-2}$ cm.

Capillary columns, according to theory, could yield values of H between one-quarter and one diameter, depending upon the amount of sorption, that is, retention. Thus, if we settle on the value, one-half a tube diameter, we are likely to be close to a realistic estimate. The narrowest practicable column is about 0·254 mm., that means that the least value of $H_{min.}$ is about $1{\cdot}2 \times 10^{-2}$ cm. We thus see that there is little to choose between packed and open columns on the basis of H.

In practice, the best values of $H_{min.}$ realized with conventional packed columns are close to 3×10^{-2} cm., that is, about twice the best attainable, although Halasz and Heine (1968) have quoted an aerogel, gas–solid column as having $H_{min.} = 1 \cdot 2 \times 10^{-2}$ cm., close to the above quoted best attainable. Capillaries too, have been found to come within about a factor of two of theoretical best. Thus we see that there is very little of practical value to be gained by further research in this area and it follows that the means to achieve excellent efficiencies are already well established. They are to be found described in any modern text.

Studies in column selectivity

Although the literature probably contains thousands of papers nominally dealing with the subject of selectivity, these are almost all of the type in which the relative merits of solvents or adsorbents A, B and C in the separation of mixture X, Y and Z are compared. Rarely, if ever, are we thus provided with the general information we need to extrapolate our experience to other systems.

The normal gas liquid chromatography column relies for its selectivity on differences in partition coefficient of sample components between the solvent and the gas phase. Most solutions deviate considerably from ideal solution behaviour but, unfortunately, molecules of similar type usually deviate in the same direction and to similar extents. Thus, frequently, we find that the degree of separation observed is not dissimilar from that corresponding to ideal behaviour. The most readily achieved enhancement of separation due to sorption forces is that between types. This can often be quite remarkably large but is then occasionally accompanied by undesirable side effects, in particular, peak asymmetry.

It would be of great advantage to be able, if only roughly, to calculate values of α on a theoretical basis. Theories of solution are legion but few are in any practical sense successful. A combination of the Hildebrand-Scatchard and Flory-Huggins approaches has been found to be quite satisfactory as a description of a number of gas liquid chromatography systems. However, these theories apply only to systems in which molecular interactions are relatively weak and so, at this time, they are of very restricted value for analysts. The prospects are, nevertheless, quite good, since gas liquid chromatography has established itself as a competitive means for the study of solutions and there is good reason to expect considerable theoretical advance as a consequence. The subject has been placed in perspective recently in a review by Langer and Sheehan (1968) and they illustrate a number of interesting possibilities for solvent choice and the "tailoring" of solvent molecules for specific separations.

Of more immediate interest for those searching for high selectivity are the novel areas of organic complexing systems in gas liquid chromatography and salt-modified adsorbents in gas solid chromatography.

Quite spectacular improvements in separation of a wide variety of solute types have been achieved by the use of transition metal complexes (for example, Barber *et al.*, 1959). Fig. 2 illustrates the relative separation of a wide range

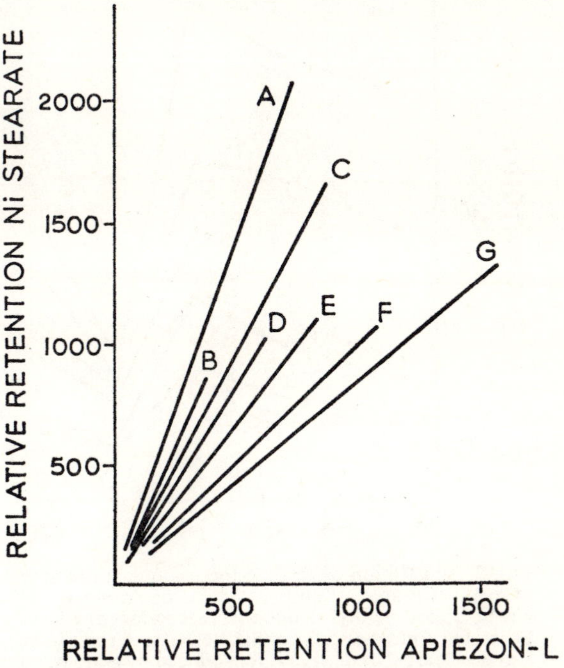

FIG. 2. Relative retentions (mesitylene as standard with $V_R' = 1000$ ml.) on nickel stearate and Apiezon-L at 156° C. A: *n*-primary alcohols; B: 2-methyl alcohols; C: secondary alcohols; D: cyclic ketones; E: aliphatic ketones; F: aromatic hydrocarbons; G: aliphatic hydrocarbons (Barber *et al.*, 1959; reproduced by permission of *J. chem. Soc.*).

of solute types achieved with nickel stearate as opposed to Apiezon-L. The greater selectivity of the stearate columns is indicated by the fanning out of the plots and the fact that the slopes vary considerably. Fig. 3 compares the several transition metals in this context and it is seen that a wide range of effect can be achieved by appropriate choice. Complexing reactions of the charge-transfer type too can be utilized (Langer and Purnell, 1963, 1966; Cadogan and Purnell, 1968) and these are particularly useful for aromatic separations.

Gas solid chromatography adsorbents have, in general, found little favour because of the commonly observed peak asymmetry, occasional catalytic activity and reputed irreproducibility. However, gas solid chromatography

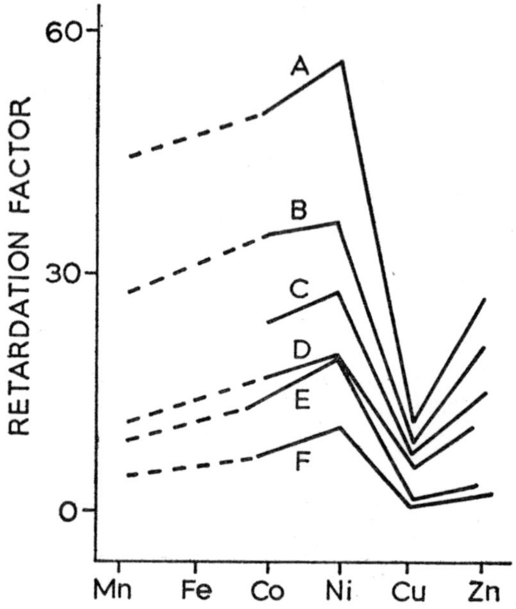

FIG. 3. Illustration of relative retention as a function of metal stearates. Elution at 156°C of A: *n*-primary alcohols; B: 2-methyl alcohols; C: secondary alcohols; D: tertiary alcohols; E: cyclic ketones; F: aliphatic ketones (Barber *et al.*, 1959; reproduced by permission of *J. chem. Soc.*).

systems offer very remarkable specificity in many cases and the permanent gases can only be handled in this way. The development of salt-modified adsorbents by Phillips and Scott (1968) has exciting possibilities since most of the defects of gas solid chromatography are overcome while not only are very

TABLE I

VALUES OF $N_{req.}$ AS A FUNCTION OF α AT $k \geqslant 10$

α	$N_{req.}$
1·05	16 000
1·50	325
2·50	100
5·0	56
10·0	45

selective separations possible but the specificity can be varied quite remarkably by adjustment of the proportions of salt to adsorbent or the identity of the salt. Some indication of what may be achieved is given in the data of Table I.

The great advantages of these newer approaches are that they are more flexible than the single phase method. It is to be hoped that work along these lines will be energetically prosecuted in the near future.

Eliminating the column

The very superficial account up to this point will, it is hoped, have at least established with reasonable conviction the view that the most profitable route for study is that designed to improve column selectivity. Fig. 4 shows something of the importance of this factor. Figs. 4*a* and 4*b* are taken from the work of Desty, Goldup and Swanton (1959) and show a comparison of chromatograms obtained for an aromatic mixture with a 120-cm. packed column and a 14-m. capillary, both coated with 7,8-benzoquinoline. The theoretical plate height achieved with the packed column is actually superior to that of the capillary but the vastly greater value of N available in the latter case yields the separation. The properties of the solvent, however, are the really important feature since it is remarkably selective for these substances, for example, $\alpha = 1 \cdot 08$ for *m*/*p*-xylene. This is brought out by Fig. 4*c* which shows a chromatogram obtained by Bohemen and Purnell (1958) with the same mixture run on a packed column, 8·5 m. long and having 50 per cent more plates than the capillary of Fig. 4*b*. Yet, because the solvent, PEG 400, is very unselective for aromatics ($\alpha = 1 \cdot 03$ for *m*/*p*-xylene) separation is not achieved.

We can put this discussion on a reasonably quantitative basis if we consider the following form of the basic separation equation:

$$N_{req.} = 36\left(\frac{\alpha}{\alpha - 1}\right)^2\left(\frac{1 + k}{k}\right)^2 \tag{4}$$

Here, $N_{req.}$ represents the number of plates required of a column if it is to yield a six, standard-deviation separation of two components of separation ratio α at some value of k for the second emergent peak. k, the capacity ratio, is defined by V_{R2}/V_g, where V_g is the retention volume of "air".

Suppose we consider the simplest analysis possible, that in which we separate two substances one of which has $V_{R1} \simeq V_g$, the other having $V_{R2} \gg V_g$. This we may call the infinitely easy separation. Since α and k are very large, both $(1 + k)/k$ and $[\alpha/(\alpha - 1)] \simeq 1$, hence,

$$N_{req.} = 36 \tag{5}$$

Recalling that the smallest practically attainable $H_{min.}$ lies between 1 and

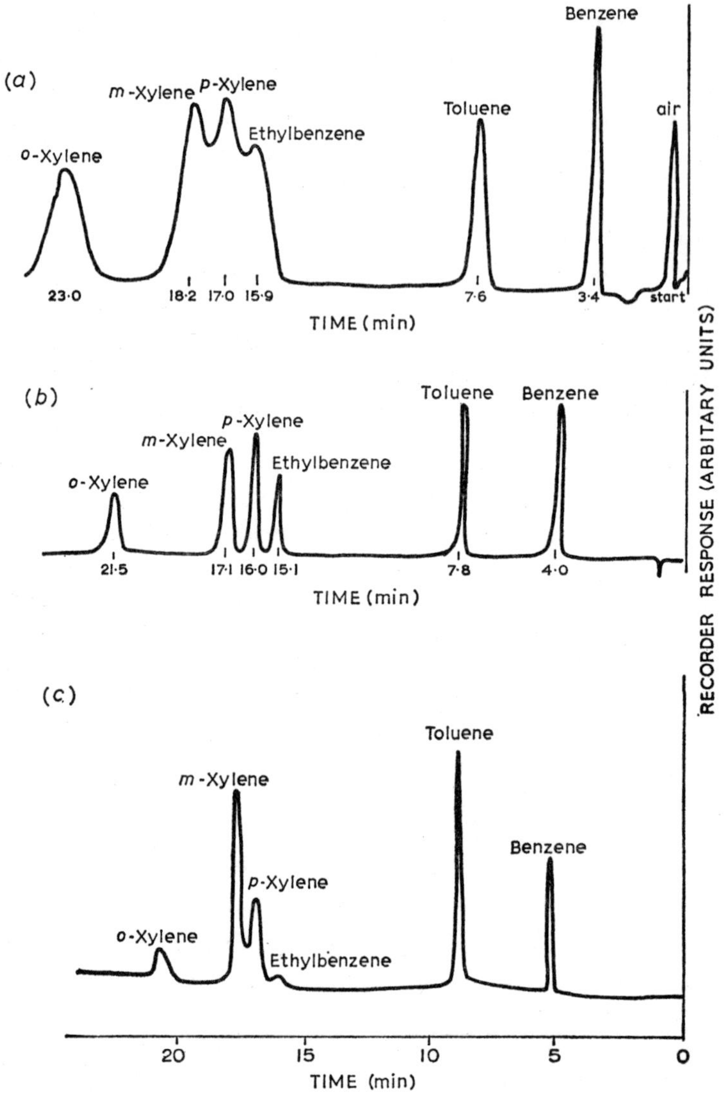

FIG. 4. Comparison of importance of N and α in analysis. Aromatic hydrocarbons run on: (*a*) 120-cm. packed column of 1600 plates ($H = 0 \cdot 075$ cm.); (*b*) 1200-cm. capillary column of 14 600 plates ($H = 0 \cdot 083$ cm.), both columns contain 7,8-benzoquinoline which gives good α; (*c*) 800-cm. packed column of 22 000 plates ($H = 0 \cdot 035$ cm.) containing PEG. 400. Note that column (*a*) is slightly more efficient than column (*b*) but has too few plates, i.e. is too short. Note also that column (*c*) is twice as efficient as either (*a*) or (*b*) and has more plates but fails because α is very unfavourable. (Chromatograms (*a*) and (*b*) from Desty, Goldup and Swanton, 1959; reproduced by permission of *Nature, Lond.*)

3×10^{-2} cm., we may thus immediately estimate that the length of the shortest possible column we might nowadays use lies between 3·0 and 1·0 mm. At the corresponding $u_{min.}$ the analysis times would then lie in the range 0·01 to 0·1 sec. Very few analyses can be reduced to 36 plate analyses since infinite α can only be achieved occasionally.

TABLE II

ILLUSTRATION OF SELECTIVITY ATTAINABLE BY VIRTUE OF CHEMICAL COMPLEXING. "SOLVENT", MANGANESE STEARATE AT 156°, COMPARED WITH APIEZON-L

Solutes	*Boiling points* (°C)	α (*Apiezon*)	$N_{req.}$ (*high k*)	α (*Stearate*)	$N_{req.}$ (*high k*)
β-Picoline	143	1·009	4×10^5	1·29	6×10^2
γ-Picoline	143				
n-C_4H_9OH	118	1·10	4×10^3	1·45	$2{\cdot}5 \times 10^2$
isoC_4H_9OH	108				

Table II shows the value of $N_{req.}$ for a range of α at high k (it may be noted that $k = 10$ is effectively $k = \infty$). We see that if we can achieve only $\alpha = 2{\cdot}5$, our limiting column length should still not exceed about 100 mm. This sort of α is well within the optimistic range of attainable selectivities for a wide range of analyses. What is really important to note, however, is how much greater is the potential gain in terms of column length of improved α than anything we can hope for from further efficiency studies.

Discussion

Few workers would agree that the aim of fundamental research in the past fifteen years has really been the elimination of the column. It is, of course, a contradiction in terms, yet the search for efficiency and selectivity must have had some purpose. In most cases this has been taken to be that of getting more separating power from a given column but, this in itself is only another way of saying that the aim is to make the column as short as possible. What are the advantages of this philosophy? There are some obvious gains such as the reduced pressure drop, the contingent easing of the sample introduction problem, often a lower column temperature becomes feasible, cost may be lowered and constructional reproducibility may be enhanced. All these are worth having. The major gain, however, is in reduced analysis time. This is clearly defined as

$$t = \frac{L}{u} = \frac{N_{req.}H}{u} \tag{5}$$

which, if we choose to work at high velocity (well above $u_{min.}$) reduces to

$$t = N_{req.}C$$

Thus, for a given column, analysis time is directly determined by N_{req}. Any reduction in column length as a result of increased α or reduced H thus offers gains in analysis speed. If we now refer again to Table I we see that on going from $\alpha = 1 \cdot 05$ to $\alpha = 2 \cdot 5$, not only do we reduce the column length required by a factor of 160, but we simultaneously reduce the analysis time by the same amount.

The validity of this argument is unquestionable and has been well demonstrated in practice (Purnell and Quinn, 1960; Scott and Hazeldean, 1960)

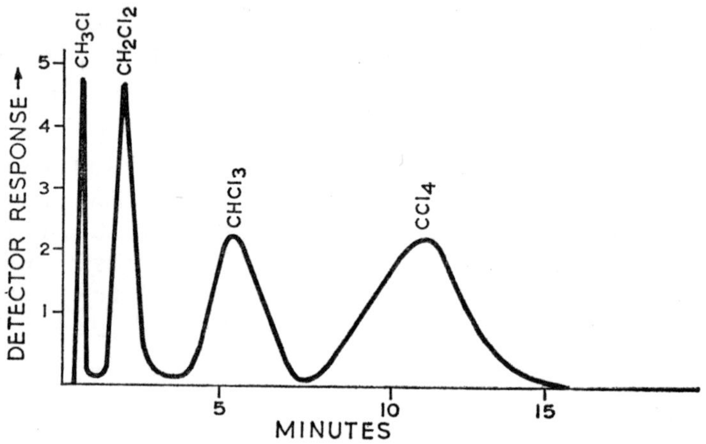

FIG. 5. Separation of chloromethanes with a 12-cm. column of liquid paraffin/stearic acid. (Purnell and Spencer, 1955; reproduced by permission of *Nature, Lond.*)

although few workers have attempted seriously to follow this path. It is perhaps a little surprising that this should be so.

This discussion has ranged widely and loosely and will have achieved its purpose if it draws attention to some of the more general features of the state of the art. It is perhaps fitting to conclude with a little confirmatory evidence of the proposition that columns of a few centimeters can do the job, given a favourable α. Fig. 5 illustrates the separation of the chloromethanes on a column less than 13 cm. long. It is a salutory thought that this chromatogram was first published in 1955 (Purnell and Spencer, 1955).

SUMMARY

The state of gas chromatographic theory and practice is briefly reviewed. It is suggested that column efficiencies now attainable (theoretical plate heights) are close to the best predictions of theory. In contrast, column selectivities are still little improved over those attainable a decade ago. The greater importance of selectivity than of theoretical plate heights in determining column dimensions is illustrated. Emphasis is given to the need for further basic research on selectivity since it is proposed that an appropriate chromatographic philosophy is the reduction of column length to a practical minimum. Examples of promising new areas are given.

REFERENCES

BARBER, D. W., PHILLIPS, C. S. G., TUSA, G. F., and VERDIN, A. (1959). *J. chem. Soc.*, 18–24.

BOHEMEN, J., and PURNELL, J. H. (1958). In *Gas Chromatography*, pp. 6–22, ed. Desty, D. H. London: Butterworth.

CADOGAN, D., and PURNELL, J. H. (1968). *J. chem. Soc.* **A.**, 2133–2137.

DESTY, D. H., GOLDUP, A., and SWANTON, W. T. (1959). *Nature, Lond.*, **183**, 107–108.

HALASZ, I., and HEINE, E. (1968). In *Progress in Gas Chromatography*, pp. 153–208, ed. Purnell, J. H. New York: Wiley-Interscience.

LANGER, S. H., and PURNELL, J. H. (1963). *J. phys. Chem., Wash.*, **67**, 263–270.

LANGER, S. H., and PURNELL, J. H. (1966). *J. phys. Chem., Wash.*, **70**, 904–909.

LANGER, S. H., and SHEEHAN, R. J. (1968). In *Progress in Gas Chromatography*, pp. 289–324, ed. Purnell, J. H. New York: Wiley-Interscience.

PHILLIPS, C. S. G., and SCOTT, C. G. (1968). In *Progress in Gas Chromatography*, pp. 121–152, ed. Purnell, J. H. New York: Wiley-Interscience.

PURNELL, J. H., and QUINN, C. P. (1960). In *Gas Chromatography*, pp. 184–198, ed. Scott, R. P. W. London: Butterworth.

PURNELL, J. H., and SPENCER, M. S. (1955). *Nature, Lond.*, **175**, 988–989.

SCOTT, R. P. W., and HAZELDEAN, G. S. F. (1960). In *Gas Chromatography*, pp. 144–158, ed. Scott, R. P. W. London: Butterworth.

DISCUSSION

Payne: You have been both lucid and provocative; will anyone now defend the manufacturers?

Scott: I shall not defend the manufacturers because I agree entirely with what Professor Purnell has said as far as the separation of simple mixtures is concerned. But in practice we are dealing mainly with multicomponent mixtures containing hundreds of individual substances. In considering such mixtures there are two basic types of intermolecular forces that can be exploited—London dispersion forces and polar forces (assuming there is no interaction in the gas phase). Carrying out the separation on two columns, exploiting these different forces in turn merely reshuffles the peaks from one elution order to another. Thus, using polar forces, one pair of components may be separated and another pair merged; using non-polar forces, the merged pair may be separated but the first pair is again unresolved. The philosophy of an ideal, effective column applies to simple mixtures. Separation of multicomponent mixtures, in our experience, is best achieved using long columns and two extreme types of stationary phase.

Purnell: Of course we have to deal with multicomponent mixtures all the time. My philosophy about the short column embodies the idea that we should not be thinking in terms of trying to separate all the components of a mixture in one operation but rather in a series of small, two-component perhaps, separations. A system, the principle of which I like, is one in which a sample of n components comes into a manifold of $n - 1$ columns, in each of which it is split into two. It is not too unreasonable to visualize as many as 19 columns, each (say) 10 mm. in length, in a grid. Building parallel units of many columns is a reasonable idea provided the columns are not very long; commercial attempts to build parallel systems have failed essentially because the columns have been too long to balance for retention. The main problem is how to cope with the $(n - 1)$ read-outs. A computer would probably be needed for this. With sufficient column-specificity ranges available in the unit the components could be shuffled about so much in terms of retention that one could hope for a detailed analysis of all components. The effort required to develop such an instrument would certainly be no greater than that which has already gone into some chromatographic projects.

Martin: I agree with almost everything you have said but not with your conclusions. It is the engineering side of the work that has been so staggeringly lacking. It has always been obvious that we needed several columns and wanted to use several columns on different parts of the same run. The reason that we have not done this is that nobody has yet produced a suitable piece of apparatus. I find it inconceivable that 19 or 20 stationary phases can be found that are able to separate the parts of a complicated mixture into an identifiable set of groups. If one had an elaborate enough computer programme and complete details about the components of the mixture at the start one might possibly work out something along these lines, but this is not a practical proposition in the general case. We need to complicate the *machine* so that it becomes feasible to re-run on different columns almost all the components that have been run on an original few columns. Your point about selectivity comes in here, but what we are interested in finding are columns in which the solution is ideal for one set and departs from ideal for other sets of components. I cannot eliminate the column. Gas chromatography is valuable just because there is a column that can handle many different substances simultaneously. Similar substances inevitably demand long columns.

Purnell: May I in defence say that I think the only good thing in a chromatograph *is* the column: it does all the work most economically. No matter how one misuses the column it never fails to produce a chromatogram; indeed, the chromatograph is probably the most idiot-proof machine ever invented. I want, of course, to eliminate not the column but the widespread concept that we should try to deal with ever more complicated mixtures with longer columns. On the contrary, our aim should be to keep the column elegantly simple by working with the shortest length possible.

Scott: This system will operate satisfactorily provided one is dealing with substances that can be separated using polar forces only. Taking the extreme case of hydrocarbons, which are non-polar, the only forces by which they can be separated are London dispersion forces and so roughly the same relative retention ratio will be obtained on any type of stationary phase. However, if a mixture of different solute types is incompletely separated on one stationary phase then a change to another may provide an improved separation. I would like to repeat that in our experience, for multicomponent mixtures, separations are best obtained using long columns in two extreme types of stationary phases.

Concerning column efficiency, it is very difficult to achieve more than 3000 plates/metre for any given column system. I have never achieved more than this, but the difference between 3000, say, and 600 plates/metre (which is the efficiency one sees from the majority of columns used) is considerable. It is relatively easy to obtain 2000 plates/metre from a column and this produces a tremendous improvement in resolution compared with the same column when it has only 600 plates/metre.

Purnell: I agree with you. We commonly get about 2000 plates/metre and my colleagues say that it is not worth struggling for more.

Carter: As a manufacturer I agree with what has been said about us; we would indeed like to simplify the column and we do have to compromise in our designs because we do not know what sort of column will be fitted by the user. But users seek simplicity as well as efficiency; six columns would satisfy about 90 per cent of the market. The snag with highly specific columns is that until an analysis has been made the column user does not know which specific column he wants; and he may want a different column for each analysis. This means having many columns available whereas, ideally, the user would like one universal column that he can simply put into his instrument and leave there. A column with far too many theoretical plates would probably be preferred by most users, so that they need not worry about whether they are using the best stationary phase, just as one might prefer a computer which has a large core so that too much time is not spent on preparing an optimum programme. We are interested in the selective columns which use salt-modified adsorbents (Scott and Phillips, 1965). We have shown some of these at exhibitions and circulated literature on them without one single sale as a result. Nevertheless, we are sufficiently convinced of the potential merit of selective columns to continue our research. We are trying to develop an instrument with the shortest column possible for useful measurement of retention data, even though we do not expect a direct commercial outlet for such a machine. A battery of very short columns might help in selecting the most appropriate specific column even if it did not do the actual analysis or identification.

Purnell: I am not surprised that no one wants to buy these selective columns (see also Phillips and Scott, 1968); this field of study is foreign territory to most column users. But when we start thinking in terms of chemical complexing systems as substrates for gas liquid chromatography, an enormous range of systems becomes available for us to work with instead of just a small number of non-volatile liquids. All the many complexing reactions now known to occur in water could be looked at again in non-aqueous media as potential gas liquid chromatography systems of high selectivity. Further, if we could use water as the solvent for complexing agents in gas liquid chromatography, we might get astounding selectivities for certain kinds of systems. Water, unfortunately, is not really available for this. However, even in non-aqueous media a selectivity bonanza may be awaiting us!

Scott: But water is available as a stationary phase.

Purnell: It is not a helpful stationary phase in practice.

Scott: If you want to use an aqueous medium there is no reason for not using water.

Purnell: Although water can and has been used, it does not appeal to the average user because of the complications—presaturation for example—that accompany its use.

Scott: These are nothing compared to the complications of the rest of the hardware.

Purnell: Maybe, but one significant complication is that water is not a good solvent for organic materials.

Carter: Different sorts of research are involved here. A manufacturer can be expected to try and improve the quality of his columns. But to work on new column materials is a different matter. This would have scientific but not commercial value.

Brooks: Surely plenty of different types of stationary phase are available. Selectivity has been extensively studied, especially for the "biological" types of molecules—steroids and similar compounds. In this field there remains what seems to the chemist to be a simple engineering problem: how can we get samples on to the column without having to use vast quantities of a solvent in which we are not interested? This elementary step has been studied relatively little. We do not need very short columns except for particular purposes. A switching arrangement whereby we could use three or four different stationary phases of conventional type and length would solve many problems.

I agree with you, Dr. Scott, that when dealing with mixtures containing many unknown components one needs a very efficient column; but if one is handling mixtures of known composition, chemistry alone can often compensate for lack of column efficiency.

Martin: In gas liquid chromatography, particularly with proteins, although the opportunity to use both highly selective and totally non-selective phases exists, Porter and I (Martin and Porter, 1951) as the originators of the method have been almost the only users of the highly selective phases, probably because their preparation is so tedious. It is much easier to use a Sephadex column, which is known to work, even if not too efficiently, than to spend several weeks trying to find a system suitable for a particular separation. Similarly, if highly selective phases for other purposes are ever produced people probably will not use them.

Purnell: I agree. People do not generally use highly efficient columns; there has been no progress in the development and use of high-efficiency columns for some years. Most column users are happy to operate at a theoretical plate height of around 1 mm. because such relatively inefficient machines do the job all right. This is not a desirable state of affairs. However, in liquid–liquid chromatography it *is* essential to keep the column short.

Lipsky: Pellicular liquid–solid systems have one great advantage: for the first time, by changing the nature of the eluent as with gradient elution, analysis time can be greatly reduced. After the completion of a particular analysis, within periods of a few seconds to a few minutes, the solvent system can be completely changed and returned to its original concentration without disturbing the equilibrium of the column. This is because little swelling or contraction occurs with these thinly coated surfaces. The possibility of doing these sorts of experiments opens up entirely new parts of chemistry which can be made use of in the technology of separation.

Scott: In liquid chromatography we have another degree of freedom in our separating conditions as there can be selectivity in the mobile phase. In selecting phase systems in liquid chromatography we should remember that the relative forces exerted on a solute on a solid surface can be far greater than the forces on the solute in a liquid. For this reason more effective separations can usually be obtained by liquid–solid than liquid–liquid chromatography.

Sjövall: Liquid chromatography gives one the possibility for building suitable stationary phases. Compounds of high or low molecular weights can be covalently bound to polymer matrices (Silman and Katchalski, 1966; Porath, 1968). The use of stationary gel phases carrying specific substituents in the field of liquid chromatography is likely to increase.

Scott: Is the stationary phase initially solid?

Sjövall: Yes, and one can bind something chemically on it and it may form a gel in the mobile phase.

Scott: With such a system it is uncertain whether a liquid–liquid or liquid–solid distribution is present. If the system is predominantly a liquid–liquid one selectivity will be far less than it is with a liquid–solid system.

REFERENCES

MARTIN, A. J. P., and PORTER, R. R. (1951). *Biochem. J.*, **49,** 215–218.
PHILLIPS, C. S. G., and SCOTT, C. G. (1968). In *Progress in Gas Chromatography*, pp. 121–152, ed. Purnell, H. J. New York: Wiley-Interscience.
PORATH, J. (1968). *Nature, Lond.*, **218,** 834–838.
SCOTT, C. G., and PHILLIPS, C. S. G. (1965). In *Gas Chromatography*, pp. 266–282, ed. Goldup, A. Amsterdam: Elsevier.
SILMAN, I. H., and KATCHALSKI, E. (1966). *A. Rev. Biochem.*, **35,** 873–908.

DETECTORS FOR GAS CHROMATOGRAPHY

D. W. Hill

Research Department of Anaesthetics,
Royal College of Surgeons of England, London

Perhaps the most striking thing about the technique of gas chromatography is the manner in which it has diversified in application since the original papers by James and Martin (1951, 1952). Then a simple manual titration technique was employed to observe the elution of fatty acids from the column. Physical techniques have mostly been developed for detecting quantitatively the emergence of peaks from columns, since an electrical output for a recorder is easily provided. So many techniques have been tried that over a hundred references are available covering the years 1956–1968 and the history of detectors reads rather like the history of physics. The general aim has been to devise detectors which would have a high sensitivity which was largely gas flow independent and which could be predicted in regard to the sample structure. Ultimately, a limit to the usable sensitivity is set by the level resulting from base signal instability ("noise"), loss of stationary phase from the column ("column bleed") and possible contamination from piping. These limitations naturally become more acute with high-temperature column operation and programmed temperature working.

Most manufacturers of gas chromatographs for laboratory use now concentrate on providing modular instruments which the user can adapt to suit his own requirements. A choice of three detectors is commonly provided: thermal conductivity, flame ionization and electron capture. Conditions can usually be arranged so that the response of the detectors is sufficiently linear for a voltage-to-frequency type of integrator to be used in conjunction with a monitor recorder. A number of specific detection systems such as the flame emission detector, the microwave absorption detector and the electrolytic conduction detector have emerged in recent years. These devices give a high sensitivity for specific types of compound.

An increasing use for gas chromatography is in the analysis of complex multicomponent organic mixtures, for example in the petroleum and essential oil industries and in biochemistry. Here, the column is used as an efficient means of separating the sample constituents while the detector records their emergence from the column. By means of refrigerated traps the contents of a particular peak can be isolated and identified with the aid of an infra-red (i.r.) spectrometer, mass spectrometer or nuclear magnetic resonance (n.m.r.) spectrometer. Whilst retention volume studies can identify a number of compounds, the detector is now mainly used for for quantitative purposes, sample component identification being subsequently performed on other apparatus.

THERMAL CONDUCTIVITY DETECTORS (KATHAROMETERS)

Neglecting column bleed and possible contamination, the chromatographic column elutes either pure carrier gas or (carrier gas + a sample peak). Thus the presence of a peak in the column effluent can be sensed by monitoring some physical property of the gas stream from the column. The gas density balance of James and Martin (1956) was capable of quantitative calibration but manufacturing difficulties have limited its use. One of the earliest, and still popular, detection systems is based on the recording of thermal conductivity changes in the gas, using a simple thermal conductivity cell or katharometer. This consists of a metal block drilled with separate channels through which the column effluent and pure carrier gas flow. Each channel carries an axial wire of tungsten or tungsten–rhenium. These wires are connected in opposite arms of a d.c. Wheatstone bridge circuit fed from a constant current supply, a typical bridge current being 250 mA. The fact that the wires are situated in opposite arms of the bridge provides a measure of compensation against changes in the wire resistance arising from ambient temperature changes. When a peak passes one wire, the thermal conductivity of the gas stream on that side differs from that on the other and the bridge suffers a small un-balance. This may be only one part in several thousand of the resistance of the wire. The voltage output from the bridge is fed via an adjustable attenuator to a one millivolt potentiometric recorder.

Apart from limited sensitivity, baseline drift is the big problem with katharometers. The detector depends upon measuring a small change in resistance of a much larger resistance whose value is dependent on gas flow, bridge current and ambient temperature. A stable power supply is vital, with a well-lagged block and steady gas flows. The sensitivity of the detector depends on the difference in temperature of the wire and the block and on the difference in the thermal conductivities of the gases surrounding the wires. Hydrogen or helium are generally used as the carrier gases since their thermal conductivities are substantially greater than those of most other gases and vapours. The presence of a sample peak in one side of the detector will lower the effective thermal conduction on that side and increase the wire resistance. This is arranged to generate a positive peak on the recorder chart. Negative peaks arise with some carrier gases when the presence of a peak increases the effective conductivity. The effect of the choice of carrier gas on the sensitivity of a katharometer is discussed by Ray (1958), Eden, Karmen and Stephenson (1959), Pecsok and Windsor (1968) and Castello and D'Amato (1968).

Thermal conductivity detectors are still widely used, particularly for the analysis of light gases. Forsey (1968) used parallel columns of molecular sieve and polymer beads with a katharometer to analyse hydrogen, oxygen, nitrogen, methane, carbon monoxide and dioxide, ether and water, and a similar arrangement can be used for the analysis of gases in blood (Hill, 1966). Indeed, the analysis of a synthetic Martian atmosphere with a thermal conductivity gas chromatograph is described by Whilhite and Hollis (1968).

Thermistor beads have been used as an alternative to hot wires in katharometers. Although capable of giving only a small dead volume, thermistors are not too stable as wires both in terms of calibration and baseline. In addition, the semiconductor bead can be affected by hydrogen carrier gas unless it is protected by a thin glass covering (Conlan and Szonntagh, 1968). A typical thermistor resistance would be 2000 Ω at a temperature of 20° C. Adlard and Hill (1960) used a thermistor gas chromatograph to analyse anaesthetic gas and vapour mixtures.

Bridge configurations designed to minimize baseline drifts have been described by Littlewood (1960) for hot wires and by Buhl (1968) for thermistors.

IONIZATION DETECTORS

General considerations

Ideally, the output voltage from a detector should be sensibly zero until it receives a peak. Thus there is no need to keep a bridge balanced in between peaks as with the katharometer. A substantial step forward in the development of high sensitivity detectors was the argon ionization detector of Lovelock (1958*a*). In this detector, electrons from a radioactive source (strontium-90 or tritium) are accelerated by a high voltage to excite metastable levels in the argon. The metastables decay, giving rise to ultraviolet (u.v.) photons which can ionize organic molecules introduced into the detector from the column. Fig. 1 shows a cross-section of one form of commercial argon detector. The cathode comprises the ^{90}Sr foil source, and the central anode is supported from a sparking plug insulator. The d.c. polarizing voltage lies in the range 1–1·5 kV and the source current is of the order 1×10^{-8} A. This is measured with a simple balanced valve electrometer (Thompson 1959) or currently with a hybrid electrometer valve-amplifier circuit. Because of its low background current and high sensitivity the argon detector was initially very successful. But because of its marked non-linearity and sensitivity to water vapour it has largely been superseded by the flame ionization

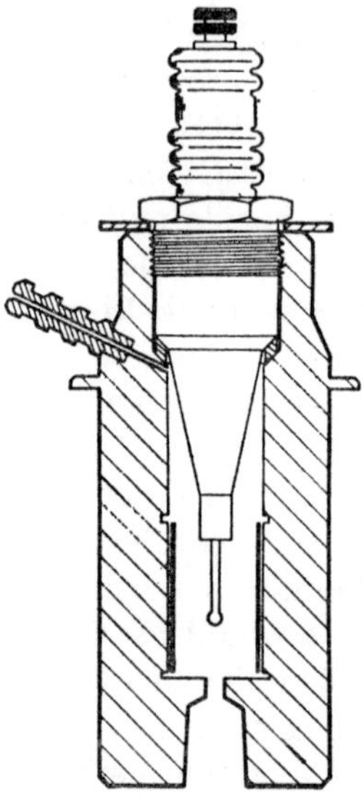

FIG. 1. Cross-section of a commercial macro-argon ionization detector (by courtesy of Pye-Unicam. Ltd.).

detector. The detector in Fig. 1 is known as a macro-argon ionization detector because it is intended for use with conventional packed columns. A micro-argon detector has been developed by Dr. J. E. Lovelock for use with open tubular (capillary) columns. In a detailed examination, Fowlis, Maggs and Scott (1964) showed that the argon detector was approximately 1000 times as sensitive as a flame detector, but was linear over only a very limited range. Lovelock had recommended the use of a high value linearizing resistor in series with the anode to limit non-linearity effects arising from space charge build-up, but Fowlis, Maggs and Scott (1964) found that the linearizing resistor did not improve the linearity.

The argon ionization detector is attractive in that it only needs the use of a single non-explosive carrier gas, but changes in the water vapour concentration of the argon can markedly effect the sensitivity. Hill and Newell (1965*a*) showed that an increase in the water vapour concentration of the argon from

8–10 ppm reduced the response to a concentration of 6 ppm of diethyl ether vapour by 20 per cent.

Hill (1962) has shown that the range of the primary beta particles is small compared with the dimensions of an argon detector, 90 per cent of the ionization occurring within 0·4 mm. of the foil. As well as producing ions, the beta particles will form excited argon atoms and molecules when the electric field strength is sufficient for them to gain enough energy. The rare gases are characterized by the presence of high energy resonance levels with metastable states adjacent within a fraction of an electron volt. For an argon atom, the resonance levels are 11·62 and 11·83 eV with metastable levels at 11·54 and 11·72 eV. These can ionize most organic vapours. The resonance levels are short-lived and decay through collisons to the atomic metastable levels (Colli and Facchini, 1952). In pure argon, the atomic metastable states have two predominant processes of decay, either two-body collisions to the ground state or three-body collisions to form metastable molecules. The existence of three-body collisions for the decay of the argon metastable atom was proposed by Phelps and Molnar (1953). The three-body decay to the metastable molecular state is over 100 times more probable than the two-body decay. At 101 kN/m.^{2} (760 torr) the mean lifetime of the metastable atom is 0·19 μsec. compared with the value of 3·4 μsec. found for the metastable molecule. The metastable molecule eventually decays by the emission of a u.v. photon of approximately 10 eV energy (Colli, 1954). This conclusion has been verified by the careful work of Bennett (1968) who found in pure argon, at atmospheric pressure under argon-detector conditions, photons containing only 10 eV. This confirms that it is the metastable molecule which is responsible for ionizing the sample in an argon detector. The metastable molecules exist where they are formed; diffusion losses to the walls at atmospheric pressure are negligible. Some free electrons might acquire the possibility of ionizing sample molecules in competition with the metastable molecules (Shahin and Lipsky, 1963*a*, *b*). Although this effect will result in the ionization of organic sample molecules, it will also result in quenching the production of metastable levels. The cross-section for this type of ionization is probably low, so that ionization by electrons will only be important at high organic concentrations when a reduction will be manifest in the increase of ionization current (Fig. 2).

Contrary to expectation, the sensitivity of argon detectors is also affected by gases such as carbon dioxide and nitrous oxide which have ionization potentials above the metastable levels for both the atomic and molecular argon (Hill and Newell, 1964). At low field strengths the sensitivity is likely to be increased, whilst it is diminished at normal operating voltages.

Lovelock (1958*b*) produced an ingenious micro-argon detector for use with open tubular columns. The anode is recessed and fed directly from the column. An auxiliary "purge" flow (carrier gas stream) of argon entering through a gauze diffuser confines the sample to the vicinity of the anode. But

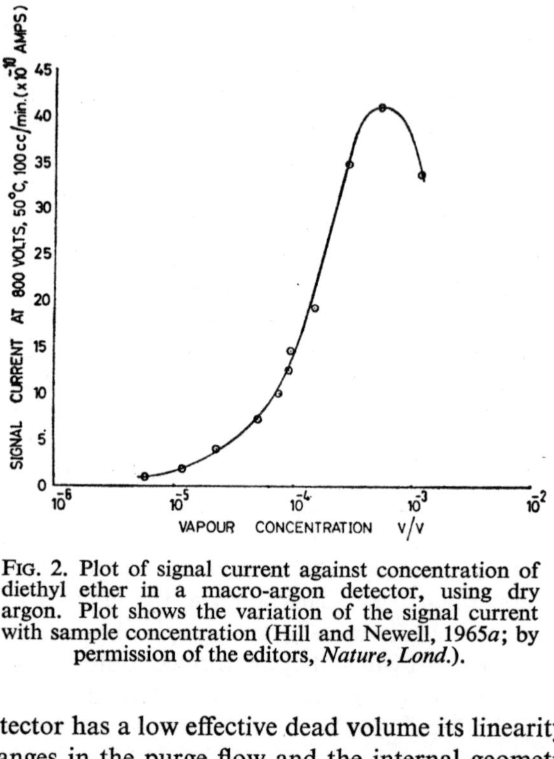

FIG. 2. Plot of signal current against concentration of diethyl ether in a macro-argon detector, using dry argon. Plot shows the variation of the signal current with sample concentration (Hill and Newell, 1965*a*; by permission of the editors, *Nature, Lond.*).

because the detector has a low effective dead volume its linearity is markedly affected by changes in the purge flow and the internal geometry. Using an electrolytic trough, Collinson, Bennett and Hill (1965) plotted the equipotentials for such a detector (Fig. 3). From this and Fig. 4 it can be seen that the region of high electric field is small and critically close to the anode. If the purge gas contains water, this can affect the emission from a radioactive (tritium) foil source in the micro detector (Warren, this volume, pp. 59–61).

The use of ionization detectors for the detection of permanent gases

Willis (1959) described the addition of 100 ppm of ethylene to the argon feed or an argon detector to produce a standing current. Oxygen, nitrogen and other gases then diminish the current, giving negative peaks. A more

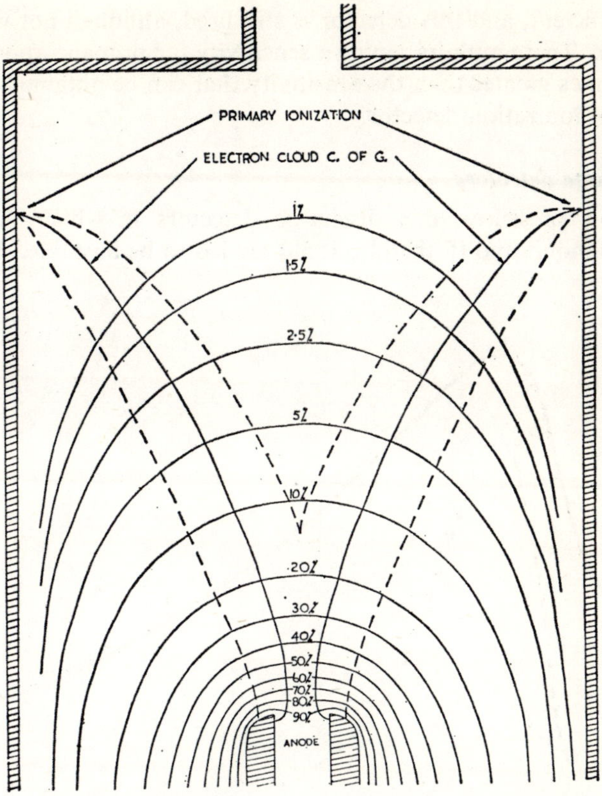

FIG. 3. Electric field distribution inside a micro-argon ionization detector. C. of G. = centre of gravity (Collinson, Bennett and Hill, 1965; by permission of the editors, *Br. J. appl. Phys.*).

elegant arrangement is to use helium which has a high energy metastable level, 19·8 eV, which can ionize all the gases and vapours. As described by Berry (1960), great care had to be taken to purify commercial helium, but it has now been shown that the passage of a high flow rate of commercial helium through active molecular sieve is adequate. The helium detector is still used for the trace analysis of permanent gases (Lansdowne and Lipsky, 1961; Parkinson and Wilson, 1968).

The photoionization detector

Basically, this uses a low pressure discharge in hydrogen with a hollow cathode arrangement to ionize the sample (Lovelock, 1960). A simple water pump is sufficient, and this detector is still used, although not widely (Price *et al.*, 1968). These authors report a sensitivity for propane approximately a thousand times greater than the sensitivity that can be obtained for propane with a flame ionization detector.

Glow discharge detectors

Early, low pressures, d.c. discharge detectors in which the discharge current was monitored (Pitkethly, 1958) tended to be unstable. Braman and

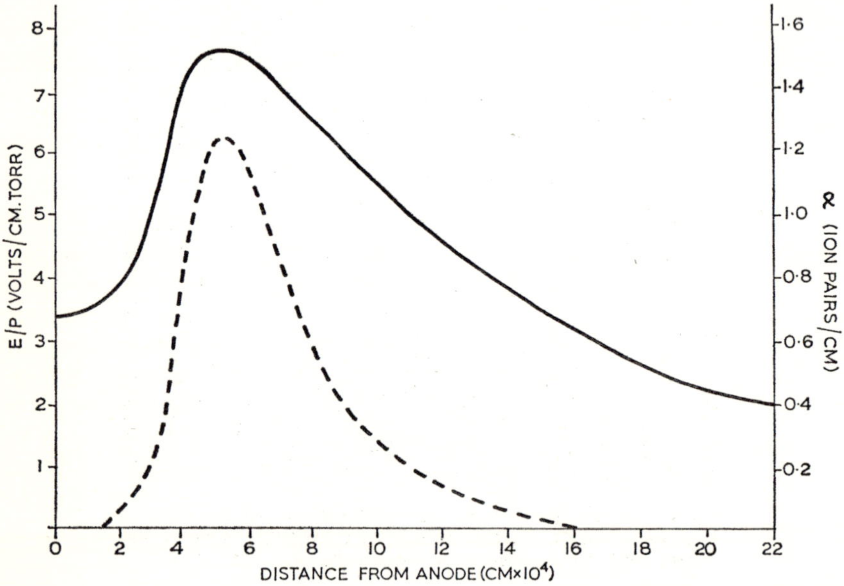

FIG. 4. Variation of the ionization coefficient, α, for electrons with distance from the anode of a micro-argon ionization detector (Collinson, Bennett and Hill, 1965; by permission of the editors, *Br. J. appl. Phys.*).

Dynako (1968) used a d.c. discharge in helium with optical filters to select the atomic emission corresponding to fluorine, bromine, chlorine and iodine.

Radio frequency detectors

In one arrangement (Williams and Winefordner, 1968) the effluent from the column flows between the plates of a capacitor which is part of the tuning circuit of a radio frequency oscillator; changes in the beat frequency between it and a reference oscillator are recorded. The detector has a sensitivity of a few parts per million for permanent gases with linear ranges of the order of 1000 ppm. Arnikar, Rao and Karmarkar (1968) monitored the discharge current of a radio frequency electrodeless discharge, in nitrogen used as carrier gas to detect peak emergence.

The flame ionization detector

At present this is easily the most popular detector in general use. Nitrogen is normally used as the column carrier gas, and this is mixed with additional oxygen and hydrogen to form a combustible mixture which is burnt (McWilliam and Dewar, 1958) as a small flame. When a peak of organic vapour is fed into the flame the ionization produced by the flame is increased and this is picked up as an increase in current across a pair of electrodes polarized with about 150 V. In most models (see Fig. 5) the insulated jet from which the flame burns is one electrode and a ring electrode is the other, although two insulated plates have been found to have advantages when used with halogenated compounds (Butler and Hill, 1961). The standing current is of the order of 10^{-11} A and the signal current about 10^{-9} A. Fowliss, Maggs and Scott (1964) showed that the flame detector could be linear for a concentration range of 10 000 : 1 and was unaffected by changes of the polarizing voltage in the range 25–150 V. Since it is self-heating, the flame detector is not so temperature sensitive as the argon detectors and is little affected by water vapour. Hill and Newell (1965*b*) found that the increase of the water vapour level in the carrier gas from 10 to 1000 ppm reduced the sensitivity to a test sample of 7 ppm (v/v) of diethyl ether by 22 per cent. The mechanism by which ionization is produced in the flame detector is uncertain. All the flammable gases—hydrogen, oxygen, and carbon monoxide and dioxide—have ionization potentials which are too high for ionization by thermal processes at flame temperatures. The response per mole of a given sample component is directly proportional to the number of carbon atoms bound only to hydrogen or to other carbon atoms in the molecule concerned. There is only a fractional contribution from those carbon atoms bound to halogens or to amines or hydroxyl groups. No contribution arises from fully oxidized carbons such as carboxyl or carbonyl carbons. On this basis it is possible to predict the relative response of the detector to various compounds.

The scope of this versatile linear detector can be greatly extended by preceding it with a small furnace containing a heated nickel catalyst at a temperature of 350–450°C. This converts the organic sample peak to methane + water, hydrogen being the carrier gas (Zlatkis and Ridgeway, 1958). These workers used the converter with C_1-C_{12} paraffins and olefins, C_5-C_8 naphthenes, and C_2-C_5 aldehydes, alcohols and ketones. (For a useful

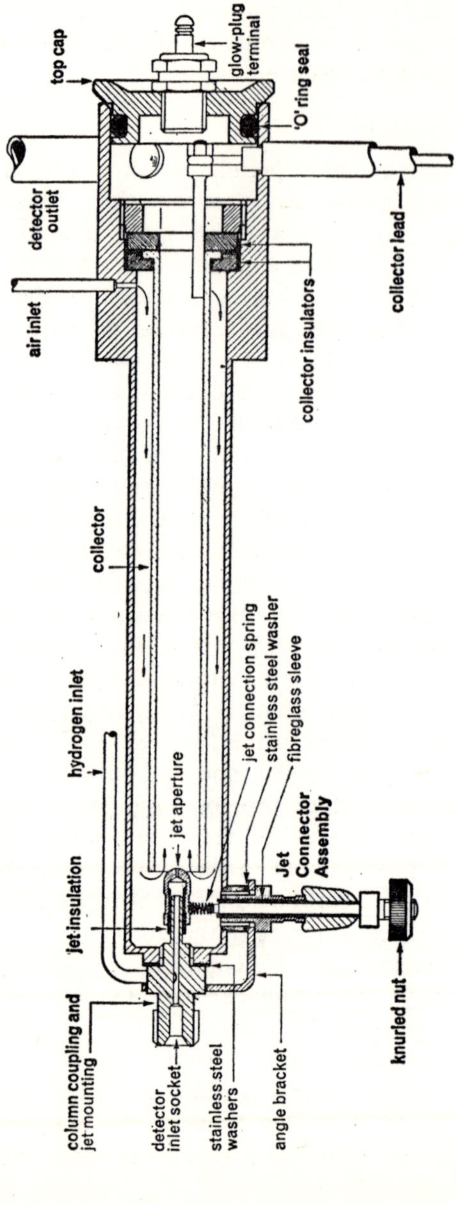

FIG. 5. Cross-section of a commercial flame ionization detector (after Maggs, 1966).

detailed evaluation of the performance of a flame ionization detector see Maggs, 1966.)

The ionization cross-section detector

It is possible to use a simple ionization chamber polarized at about 100 V as a low sensitivity detector which will respond to both gases and vapours. Strontium-90 should be used as a source of electrons whose range is substantially greater than the dimensions of the chamber (Deal *et al.*, 1956). The introduction of a sample will increase the standing current, the reduction depending upon the ionization cross-section of the molecules concerned. The current response, i, is given by the formula $i = kvQ$, where v is the polarizing voltage across the chamber, Q is the ionization cross-section of the molecules and k is a constant; and for several components by the formula

$$i = \frac{KP}{RT} \sum_j x_j \, . \, Q_j,$$

where P is the total pressure, R is the gas constant, T is the absolute temperature, x_j is the mole fraction of component j and K is a constant. Lovelock, Shoemake and Zlatkis (1964) describe a micro-cross-section detector, for use with capillary columns, which has a volume of only 8 μl.

The electron capture detector

When an ionization chamber is polarized with only about 10 V, the electrons from a radioactive source placed at the cathode will travel to the anode with little more than thermal energies. In this condition they can be readily captured by any sample molecules present which have an affinity for electrons. This particularly applies to halogenated compounds and forms the basis of the electron capture detector (Lovelock and Lipsky, 1960). As a result of electron capture the chamber current is diminished, producing negative peaks on the recorder. This simple arrangement with standing currents of the order of 10^{-9}–10^{-8} A works well with oxygen-free, dry nitrogen as the carrier gas. The sensitivity decreases by about 10 per cent for a 100°C increase in temperature. With a tritium source (100 mCi), operation is possible up to a temperature of 100°C, but with a nickel-63 source this can be extended to 350°C. Care must be taken to keep the source foil clean (Claeys and Farr, 1968), and Lovelock (1963) points out that d.c. operation may lead to the build-up of a contact potential at the cathode which can affect the chamber characteristics. Electron capture coefficients are dependent on the electron temperature so that the sensitivity is markedly voltage dependent. Finally, the calibration is likely to be non-linear. In spite of all these apparent disadvantages I have obtained useful results with a simple d.c. system. Because of their increased mobility compared with the positive ions, almost all the free electrons can be collected by applying positive pulses, in excess of an amplitude of 40 V, to the anode, the pulse width being of the order 1–5 μsec. With an interval between pulses of about 100 μsec. the electrons become essentially thermal during the field-free

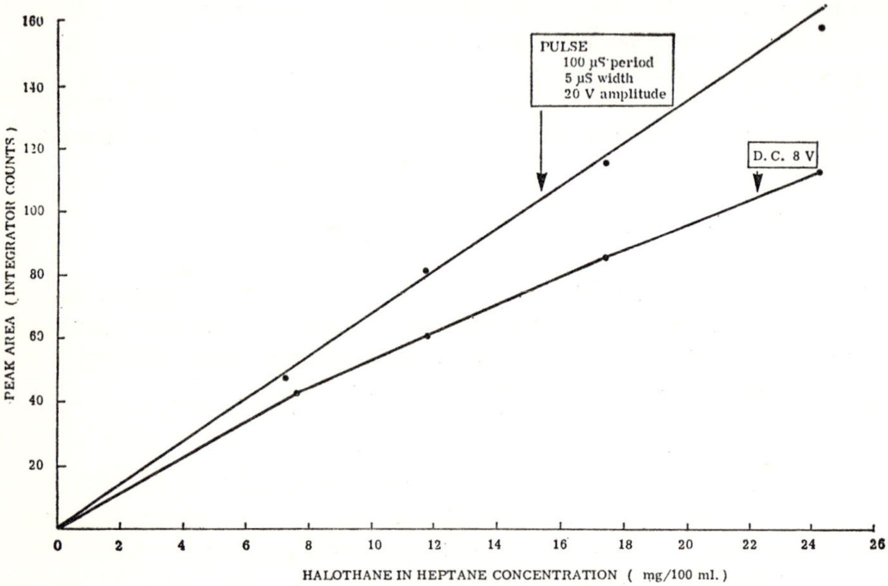

FIG. 6. Comparison of calibration curves for halothane in blood extracted with N-heptane for an electron capture detector operated in the d.c. and pulsed modes.

periods. The probability of their being captured is thus increased. The linearity of the detector can be improved by pulsing (Fig. 6) and, by choosing the optimum pulse parameters, a straight line calibration passing through the origin can be obtained (Fig. 7). With an electron capture detector, blood containing amounts of the halogenated anaesthetic halothane at levels which would produce surgical anaesthesia *in vivo* in man was diluted 1000-fold with normal heptane, and 1 μl. of the diluted blood was placed on the column (1·8 m. of 35 per cent by weight of silicone fluid MS 550 on Celite at a temperature of 45°C). Successive blood samples can be loaded at 3-min. intervals, the carrier gas being oxygen-free nitrogen. Some forms of electron capture detector (Fig. 8) are provided with an auxiliary carrier gas stream (purge flow) to dilute the sample in order to extend the range of the detector and to reduce the effective volume. Nitrogen with 1 or 2 per cent carbon dioxide, or argon with about 5 per cent methane, may be used as a quench gas in place of the purge stream (Lovelock, 1963). This increases the electron drift velocity, enabling narrow pulses to be used, and reduces the unwanted

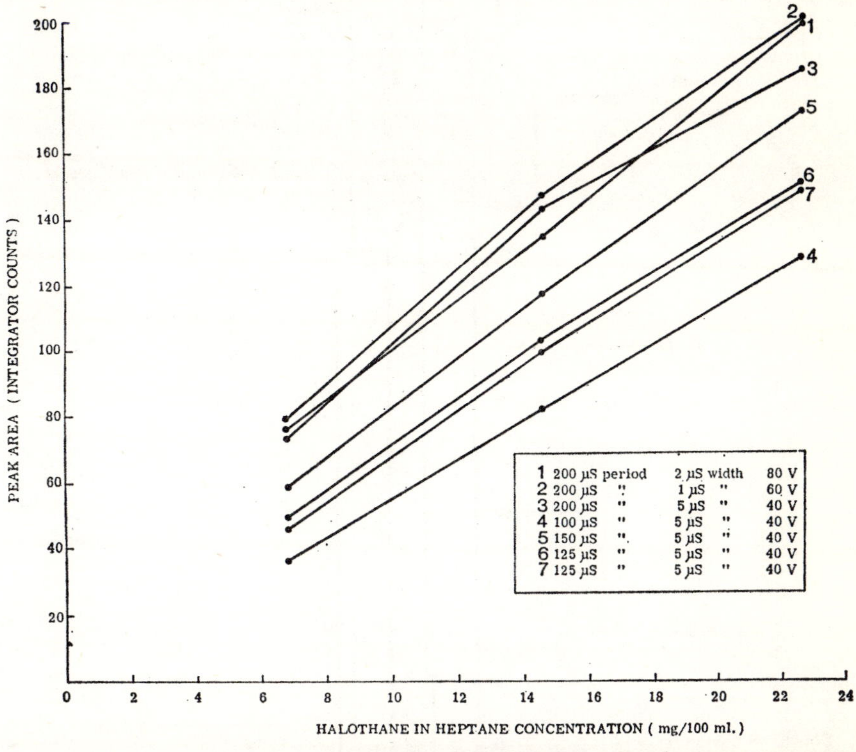

FIG. 7. Effect of changing the pulse parameters on an electron capture detector measuring halothane extracted from blood with N-heptane.

ionizing effects of metastable argon. When detecting strong electron-capturing substances we have seen no obvious benefits from the addition of carbon dioxide to the purge gas, although we found that carbon dioxide reduced the sensitivity to the electron-capturing compounds and enhanced the positive peaks recorded from non-electron-capturing sample components. Lovelock, Fenimore and Zlatkis (1967) have described the addition of a

15 MHz radio frequency field to the pulses as a means of altering the electron energies within the range 0·03–1·0 eV and thus affecting the electron capture coefficients for various substances.

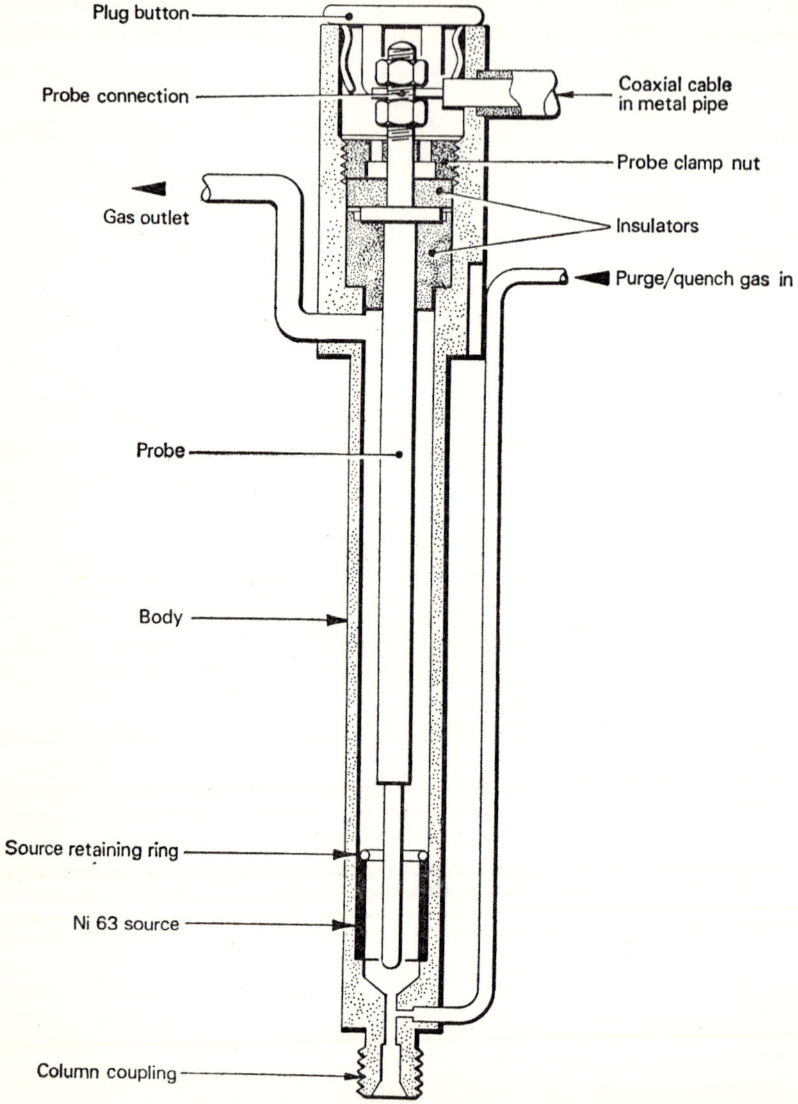

FIG. 8. Cross-section of an electron capture detector fitted with a purge or quench gas inlet (after Maggs and Swan, 1966).

SELECTIVE DETECTORS

The electrolytic conductivity detector

In this system the sample is changed into a form in which it can alter the conductivity of demineralized water, the change being recorded with a conductivity cell. Mohnke, Piringer and Tataru (1968) described the use of this detector for chromatographic analysis with an open tubular column. The electrolytic conductivity detector has also been used for the analysis of pesticides in birds and rodents. In this work it gives similar sensitivities to the electron capture detector but has the advantage that the column can be baked out with the detector in position (Coulson, 1966).

Microwave emission detector

McCormack, Tong and Cooke (1965) used a 2450 MHz electrodeless discharge in argon to excite the sample eluted from the column, and observed the emission spectrum with a half-metre scanning, grating spectrometer. The recorded spectra in most cases were traceable to diatomic radicals or free atoms. This detector can be sensitized to halogens, phosphorus, sulphur and permanent gases. The sensitivity is quoted as $2 \cdot 10^{-16}$ g./sec. for hexane.

Flame emission detector

In the flame emission detector of Braman (1966) nitrogen is used as the carrier gas, the column effluent being fed into a flame burning from a separate supply of oxygen plus hydrogen. Optical interference filters select emitted light in the following bands: 589, 515 and 414 mμ. The 589 mμ band gives the maximum emission for nitrogen-containing compounds. Nowak and Malmstadt (1968) pass the column effluent through a heated platinum spiral coated with sodium sulphate, and monitor the flame emission, using a photomultiplier fitted with optical interference filters. The introduction of halogen-containing compounds into the flame increases or decreases the emission. The arrangement is selectively sensitive to halogen-containing compounds in the sub-nanogramme range. Cochran (1966) used a sodium flame detector with phosphorus-containing pesticides. An evaluation of a selective phosphorus flame detector is given by Speakman and Waring (1968).

Semiconductor detectors

The exposure of a semiconducting film or junction to an organic vapour can alter the conductivity of the semiconductor and thus form the basis for a detector. Seiyama and co-workers (1962) used a thin film of zinc oxide with nitrogen as the carrier gas, whilst Eden and Margoninski (1968) used a reverse-biased semiconductor diode. This technique required good temperature stabilization.

CONCLUSIONS

Many alternative forms of detection have been described in the literature apart from the principle ones discussed above. (See reviews by Lovelock, 1961; Karmen, 1966; and Winefordner and Glenn 1968).

The ingenuity which has gone into the development of so many detectors is reflected in the lively progress of gas chromatography as a whole. It is now the responsibility of the user to select the most appropriate system for his requirements.

REFERENCES

ADLARD, E. R., and HILL, D. W. (1960). *Nature, Lond.*, **186,** 1045.
ARNIKAR, H. J., RAO, T. S., and KARMARKAR, K. H. (1968). *J. Chromat.*, **38,** 126–130.
BENNETT, J. R. (1968). Ph.D. Thesis, University of London.
BERRY, R. (1960). *Nature, Lond.*, **188,** 578–579.
BRAMAN, R. S. (1966). *Analyt. Chem.*, **38,** 734.
BRAMAN, R. S., and DYNAKO, A. (1968). *Analyt. Chem.*, **40,** 95.
BUHL, D. (1968). *Analyt. Chem.*, **40,** 715.
BUTLER, R. A., and HILL, D. W. (1961). *Nature, Lond.*, **189,** 488–489.
CASTELLO, G., and D'AMATO, G. (1968). *J. Chromat.*, **32,** 625.
CLAEYS, R. R., and FARR, T. (1968). *Analyt. Chem.*, **40,** 847.
COCHRAN, D. R. (1966). *Bull, Environ. Contam. Toxicol.*, **1,** 141.
COLLI, L. (1954). *Phys. Rev.*, **95,** 892.
COLLI, L., and FACCHINI, U. (1952). *Phys. Rev.*, **88,** 987.
COLLINSON, A. J. L., BENNETT, J. R., and HILL, D. W. (1965). *Br. J. appl. Phys.*, **16,** 631.
CONLAN, D. A., and SZONNTAGH, E. L. (1968). *J. Gas Chromat.*, **6,** 485.
COULSON, D. M. (1966). *Adv. Gas Chromat.*, **3,** 197–214.
DEAL, C. H., OTVOS, J. W., SMITH, V. N., and ZUCCO, P. S. (1956). *Analyt. Chem.*, **28,** 1958–1964.
EDEN, M., KARMEN, A., and STEPHENSON, J. L. (1959). *Nature, Lond.*, **183,** 1322.
EDEN, C., and MARGONINSKI, Y. (1968). *J. Gas Chromat.*, **6,** 349.
FORSEY, R. R. (1968). *J. Gas Chromat.*, **6,** 555.
FOWLIS, I. A., MAGGS, R. J., and SCOTT, R. P. W. (1964). *J. Chromat.*, **15,** 471.
HILL, D. W. (1962). Ph.D. Thesis, University of London.
HILL, D. W. (1966). In *Oxygen Measurements in Blood and Tissues*, pp. 63–80, ed. Payne, J. P., and Hill, D. W. London: Churchill.
HILL, D. W., and NEWELL, H. A. (1964). *Nature, Lond.*, **201,** 1215–1216.
HILL, D. W., and NEWELL, H. A. (1965*a*). *Nature, Lond.*, **205,** 593.
HILL, D. W., and NEWELL, H. A. (1965*b*). *Nature, Lond.*, **206,** 708–709.
JAMES, A. T., and MARTIN, A. J. P. (1951). *Biochem. J.*, **48,** PVII (abstract).
JAMES, A. T., and MARTIN, A. J. P. (1952). *Biochem. J.*, **50,** 679–690.

James, A. T., and Martin, A. J. P. (1956). *Biochem. J.*, **63,** 138–143.
Karmen, A. (1966). *Adv. Gas Chromat.*, **2,** 293–336.
Lansdowne, R. A., and Lipsky, S. R. (1961). *Nature, Lond.*, **189,** 571–572.
Littlewood, A. B. (1960). *J. scient. Instrum.*, **37,** 185.
Lovelock, J. E. (1958*a*). *J. Chromat.*, **1,** 35.
Lovelock, J. E. (1958*b*). *Nature, Lond.*, **182,** 1663–1664.
Lovelock, J. E. (1960). *Nature, Lond.*, **188,** 401.
Lovelock, J. E. (1961). *Analyt. Chem.*, **33,** 163.
Lovelock, J. E. (1963). *Analyt. Chem.*, **35,** 474.
Lovelock, J. E., Fenimore, D. C., and Zlatkis, A. (1967). In *Advances in Gas Chromatography*, pp. 188–190, ed. Zlatkis, A. Evanston, Ill.: Preston Technical Abstracts Co.
Lovelock, J. E., and Lipsky, S. R. (1960). *J. Am. chem. Soc.*, **82,** 431.
Lovelock, J. E., Shoemake, G. R., and Zlatkis, A. (1964). *Analyt. Chem.*, **36,** 1411.
McCormack, A. J., Tong, S. C., and Cooke, W. D. (1965). *Analyt. Chem.*, **37,** 1470.
McWilliam, I. G., and Dewar, R. A. (1958). In *Gas Chromatography*, pp. 142–152, ed. Desty, D. H. London: Butterworth.
Maggs, R. J. (1966). *Column*, **1,** No. 2, 2–9.
Maggs, R. J., and Swan, D. F. K. (1966). *Column*, **1,** No. 3, 2–9.
Mohnke, M., Piringer, O., and Tataru, E. (1968). *J. Gas Chromat.*, **6,** 117.
Nowak, A. V., and Malmstadt, H. V. (1968). *Analyt. Chem.*, **40,** 1108.
Parkinson, R. T., and Wilson, R. E. (1968). *J. Chromat.*, **37,** 310.
Pecsok, R. L., and Windsor, M. L. (1968). *Analyt. Chem.*, **40,** 92.
Phelps, A. V., and Molnar, J. P. (1953). *Phys. Rev.*, **89,** 1202.
Pitkethly, R. C. (1958). *Analyt. Chem.*, **30,** 1309–1314.
Price, J. G. W., Fenimore, D. C., Simmonds, P. C., and Zlatkis, A. (1968). *Analyt. Chem.*, **40,** 541.
Ray, N. H. (1958). *Nature, Lond.*, **182,** 1663.
Seiyama, T., Kato, A., Fujishi, K., and Nagatani, M. (1962). *Analyt. Chem.*, **34,** 1502.
Shahin, M. M., and Lipsky, S. R. (1963*a*). *Analyt. Chem.*, **35,** 467.
Shahin, M. M., and Lipsky, S. R. (1963*b*). *Analyt. Chem.*, **35,** 1562.
Speakman, F. P., and Waring, C. (1968). *Column*, **2,** No. 2, 2–8.
Thompson, A. E. (1959). *J. Chromat.*, **2,** 149.
Whilhite, W. F., and Hollis, O. L. (1968). *J. Gas Chromat.*, **6,** 84.
Williams, H. P., and Winefordner, J. D. (1968). *J. Gas Chromat.*, **6,** 11.
Willis, V. (1959). *Nature, Lond.*, **184,** 894.
Winefordner, J. D., and Glenn, T. H. (1968). *Adv. Gas Chromat.*, **5,** 263–300.
Zlatkis, A., and Ridgeway, J. A. (1958). *Nature, Lond.*, **182,** 130–131.

DISCUSSION

Lipsky: The chemical nature of the calibration substance used in the electron capture detector is particularly important. You are satisfied with your own particular application, but if, for example, one calibrates an instrument for certain simple halogenated materials one should not assume that the same calibration curve and degree of linear dynamic range will hold for more complex halogenated organic compounds of higher molecular weights. Other parameters, such as temperature and flow rate, must also be considered.

Scott: While we are discussing detectors I would like to describe a system I have been working on recently. This involves molecular multipliers and the idea for these came from Professor Martin. In principle, the system consists of a series of heated copper oxide and carbon units between the end of the column and, say, a katharometer detector. Eluted substances are oxidized by the copper oxide to carbon dioxide and water, and each molecule of carbon dioxide produces two molecules of carbon monoxide when it passes through each carbon stage. Thus the concentration of carbon dioxide is multiplied by two after passing through each pair of copper oxide–carbon units. The multiplier has a multiplication factor of two per stage and, for ten amplifier units, this gives a multiplication factor of 1024. As the detector is continuously detecting carbon dioxide no correction factors are necessary for quantitative analysis. We have demonstrated that the copper oxide–carbon system works with factors that agree with the theoretical conversion, and can be used to detect permanent gases such as oxygen, nitrous oxide, ammonia, hydrogen and so on. Another advantage of this system is that it can be used in conjunction with the hydrogenation method devised by Cropper (Cropper, Heinekey and Westwell, 1967), in which carbon dioxide mixed with hydrogen is passed over a heated nickel catalyst, and the methane produced is detected by a flame ionization detector having the same sensitivity for permanent gases as it does for normal organic vapours containing oxygen or hydrogen. Our system consists of a series of small furnaces, 20 mm. long and 3 mm. wide, interposed between column and detector. The copper oxide–carbon type of multiplier is not the only chemical system that can be employed. One can use an alkoxide to react with carbon dioxide to give an alcohol, or a substance such as nickel diphenyl to react with water and produce one benzene molecule for each molecule of water. Such systems could give multiplication factors of six to eight per stage.

Warren: Presumably a mixing of successive charges could occur in the multipliers. What sort of interval is needed between successive introductions?

Scott: The charges pass sequentially through the molecular multipliers; there is a constant gas flow rate and so successive charges do not interfere with each other.

Martin: It is odd that nobody has succeeded in making really practical detectors by first oxidizing organic materials to carbon dioxide and water, and then using some electrolytic method for detecting the water and a titration method for detecting the carbon dioxide. It should be quite a simple engineering project to make a detector in which each molecule of water is electrolysed away leaving one electron for every hydrogen atom in the original molecule. Practical difficulties must presumably have prevented the development of such a system for it would have a potential sensitivity some thousands of times greater than that of the hydrogen flame ionization detector.

Carter: One difficulty about such a system has been the separation of water in the sample from water in the system. We had this problem in a carbon–hydrogen–nitrogen analyser, in which we used a thermal conductivity cell to detect the water, and even with such a simple device we had trouble at first with water picked up from the connecting tubes giving "chemical noise".

Martin: What was the temperature of the detector part of the apparatus?

Carter: The furnace itself was at a temperature of about 1000° C but connecting tubes on parts of the apparatus were at about 40 or 50° C.

Scott: This is asking for trouble. I had exactly the same difficulty with the molecular multiplier but eliminated the problem by keeping all the apparatus outside the furnace at a temperature of 120° C. The production of carbon monoxide from water and hot carbon and the reduction of carbon dioxide to carbon monoxide in the carbon furnace are more complicated reactions than we might imagine. The temperature at which the conversion occurs is very critical. A change of 15° C above or below the optimum temperature results either in incomplete reduction or, at the temperature above the prescribed range the carbon becomes highly absorptive and gives rise to smeared peaks.

Purnell: What sort of time constants do these reactions have, kinetically speaking? When the samples are pushed through the tube, what is the minimum residence time for the complete conversion of carbon monoxide and carbon dioxide to carbon?

Scott: I have no firm data on the reaction rates, but complete conversion occurs if a sample contained in a gas flow of 15 ml./min. passes through a bed 10 mm. long and 2 mm. in diameter. This means that the reaction time will be a matter of milliseconds.

Purnell: Would you do anything to smear out your separations even if you had as many as ten, say, of these conversion cycles?

Scott: Not for the amplification of carbon dioxide, but the water peak would probably be smeared to some extent.

Purnell: I have never needed to analyse water so I would not worry about it in a hydrocarbon system.

Scott: In the preliminary oxidation both water and carbon dioxide are produced, and we can certainly remove the water if we do not want it. The carbon dioxide peak is slightly asymmetric after passing through the first amplifier unit and the band width may increase by as much as 50 per cent. Passage through subsequent amplifier units, however, does not seem to significantly increase either the peak width or the peak asymmetry.

Martin: These results are only preliminary.

Purnell: Of course, but this is such a beautiful idea.

Martin: But it is not in the least a new idea.

Purnell: No, it was first described many years ago, but only as a single pass device (Leipert, 1929, 1938). Provided the kinetic time constants are not unacceptable one could presumably have an infinitely long series of units—many more than 10 if one were using the middle retention region of a chromatogram.

Martin: The possibilities, using 20 units and 10^6 amplification, are quite exciting.

Scott: Our experience so far confirms that a system with 10 units is possible. The only problem appears to be that the carrier gas used must be very pure to avoid noise.

Martin: We have not yet found out the reason for all the noise in this system.

Purnell: "Bleeding" from the column may be significant and contribute to noise at this order of sensitivity.

Martin: An amplifying system such as this, in which oxygen is dissociated from copper oxide, will itself produce noise. The usefulness of this method is not yet proven.

Purnell: Before gas chromatographic techniques were available I did many of my analyses by combustion but found it a poor method, partly because the copper oxide that reached me from the manufacturer was always covered with grease. Has this been one of your problems, Dr. Scott?

Scott: We have to pretreat both the carbon and the copper oxide to clean them. At the optimum temperature for the copper oxide unit there is a slight partial pressure of oxygen over the copper oxide, and if a multiplier of a large number of these units is used this dissociated oxygen will be amplified as carbon dioxide and may produce significant noise. But there are far more efficient oxidizing agents than copper oxide.

Martin: We are looking at some other possible oxidizing agents in my department at Eindhoven.

Janák: I have two comments about your paper, Dr. Hill. The first concerns the flame ionization detector. Although this detector is well known, one parameter, the considerable pressure dependence of the efficiency of the flame ionization, has not been sufficiently discussed in the literature. We have found that the ionization efficiency (in coulombs per molecule [C/mol]), tested with traces of acetylene in the carrier gas, decreases very rapidly when the pressure at which the detector is working is increased to about 400–500 kN/m.^{2} (4–5 atm.); and increases to more than the original level when the pressure in the detector rises from 500–3040 kN/m.^{2} and higher. The relevance of this finding to analytical practice is that the marked pressure dependence of flame ionization efficiency at 760 mm. Hg may cause serious errors in the detector's response with quite small pressure changes—of the order of 10 mm. Hg.

Scott: How high is the pressure at the flame ionization detector?

Janák: About 5070 kN/m.^{2}; at this pressure the flame ionization efficiency increases steadily.

Purnell: Is the flame itself at this pressure?

Janák: Yes. The detector we have worked with is inside a metal jacket, the flame burns inside the jacket in almost the same way as in the orthodox flame ionization detector but the pressures at which the gases are fed in are controlled. The pressure regulator is attached to the outlet of the metal jacket.

Scott: How easy is it to ignite the flame in your modified flame ionization detector?

Janák: Ignition by a spark is necessary, ignition by a heated wire has not been satisfactory.

My second comment concerns a new detector (Struppe, 1963) whose sensitivity is of the same order as that of a katharometer. The function of this detector can be understood if it is visualized as a pneumatic analogue of the Wheatstone bridge. The gas corresponding to carrier gas is fed from a constant pressure source into two parallel branched tubes, both of which are equipped with two pneumatic resistances. The carrier gas from the gas chromatography column, also fed in from a constant pressure source, joins one of the branches at a point just before the second pneumatic resistance, which is itself working as the sensing element. The flow rates are controlled so that they reach a pressure equilibrium across the bridge. Any change in the composition of the column gas effluent causes a pneumatic change on the sensing element, measured as a pressure difference between zero points. This is the principal difference between this apparatus and the flow impedance detector described by Griffiths, James and Phillips (1952).

If the sensing element is a capillary, substances will be detected by changing viscosity, due to carrier gas plus solute. If the sensing element is a closed tube, the response will be a function of the molecular weight.

This device has some interesting features: (*a*) the pressure difference can be measured very precisely (in millimetres of water) at any pressure and temperature in the system of pneumatic resistances and the gas chromatography column; (*b*) the substance does not come into contact with the pressure measuring device; and (*c*) the detector works without electricity.

Hill: Has anyone in this group any experience in using a quench gas (usually a few per cent of carbon dioxide or methane in argon or nitrogen) in electron capture detectors? The function of the quench gas mixture, of course, is to suppress the levels of excited metastatic compounds when argon is the carrier gas; but I have always found that a quench gas diminishes the sensitivity of the detector.

Lipsky: We do not use a quench gas but prefer to run one gas at a time, under very carefully controlled conditions in which the operating parameters are shifted when necessary. A major problem in analysing a mixture of unknown and known substances in an electron capture detector is that a large volume of gas with a weak affinity and a small volume of gas with a strong affinity will both give the same signal. Unless one knows precisely what substances one is dealing with, the system must be constantly recalibrated.

REFERENCES

CROPPER, F. R., HEINEKEY, D. M., and WESTWELL, A. (1967). *Analyst, Lond.*, **92,** 443–449.

GRIFFITHS, J. H., JAMES, D. H., and PHILLIPS, C. S. G. (1952). *Analyst, Lond.*, **77,** 897–904.

LEIPERT, T. (1929). *Mikrochemie* (*Pregl Festschrift*), 266.

LEIPERT, T. (1938). *Michrochim. Acta*, **3,** 147.

STRUPPE, H. G. (1963). In *Gas-Chromatographie*, p. 100, ed. Struppe, H. G. Berlin: Akademie-Verlag.

THE EFFECT OF WATER VAPOUR ON ARGON IONIZATION DETECTORS

B. C. H. WARREN AND M. G. DALZELL

Physics Research Division, Chemical Defence Experimental Station, Porton Down, Wiltshire

WATER vapour is known to affect the performance of argon ionization detectors. Hill and Newell (1965) using a macro-type of detector showed quantitatively that the sensitivity towards organic substances was markedly reduced by the presence of water vapour in the carrier gas. Published information on the signal given by water vapour, however, is conflicting. MacKay (1960) observed a large negative signal but Swoboda (1960) only a slight negative signal with respect to standing current. Negative signals may be explained in terms of the electron capturing properties of water vapour (Wilkinson, 1950). Lovelock (1958), reporting on his original detector, quoted a small positive response to water vapour but observed no response with the smaller forms of detector (Lovelock, 1960).

The present communication shows that either positive or negative signals can be obtained from water vapour by varying the detector flow conditions. A positive response at high field strengths may result either from the direct ionization of water (ionization potential: 12·6 eV) through the agency of higher-energy metastable states of argon, or from complex ion formation (Shahin and Lipsky, 1963). The existence of sufficiently energetic metastable states with excitation potentials greater than 11·7 eV has been reported by several investigators (Sanders, Hurst and Bonner, 1960; Strickler and Arakawa, 1964; Hurst, Bortner and Glick, 1965). The results presented here show that the signal given by water vapour is positive if the sample is prevented from contacting the energizing source. Experiments were undertaken using a micro-argon detector constructed after the design of Lovelock (1961). At an applied potential of 1·5 kV a 200-mCi [^{3}H]titanium source gave rise to a background current of $1 \cdot 7 \times 10^{-8}$ A. Initially a steady stream of argon containing 40 ppm (v/v) of water vapour produced from a calibrated diffusion cell was admitted to the detector through the anode. The detector was maintained at a temperature of 50° C and purged with argon dried by passage

TABLE I

DIRECTION OF SIGNAL FOR DIFFERENT FLOW CONDITIONS

ANODE FLOW ml/min	PURGE FLOW ml/min 25	50	100	150
5	+	+	+	+
10	+	+	+	−
15	+	+	+	−
20	−	−	+	−
30	−	−	−	−

ANODE 1mm I.D., recessed 2mm inside well

through anhydrous magnesium perchlorate. Table I shows the direction of response using different flow combinations.

When a Strontium-90 source was used in place of the tritium source, positive signals were obtained for all flow conditions; the response curve is shown in Fig. 1. No quantitative relationship between the magnitude of response and the concentration of water vapour was established in the case of negative signals. This difference in behaviour can probably be explained by the adsorption of water vapour on the energizing source. If the flow conditions allow the sample to contact the radioactive foil and adsorption

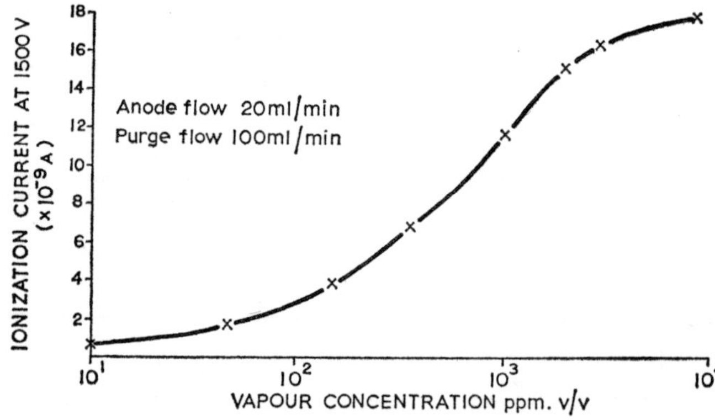

FIG. 1. Response of micro-argon detector to water vapour.

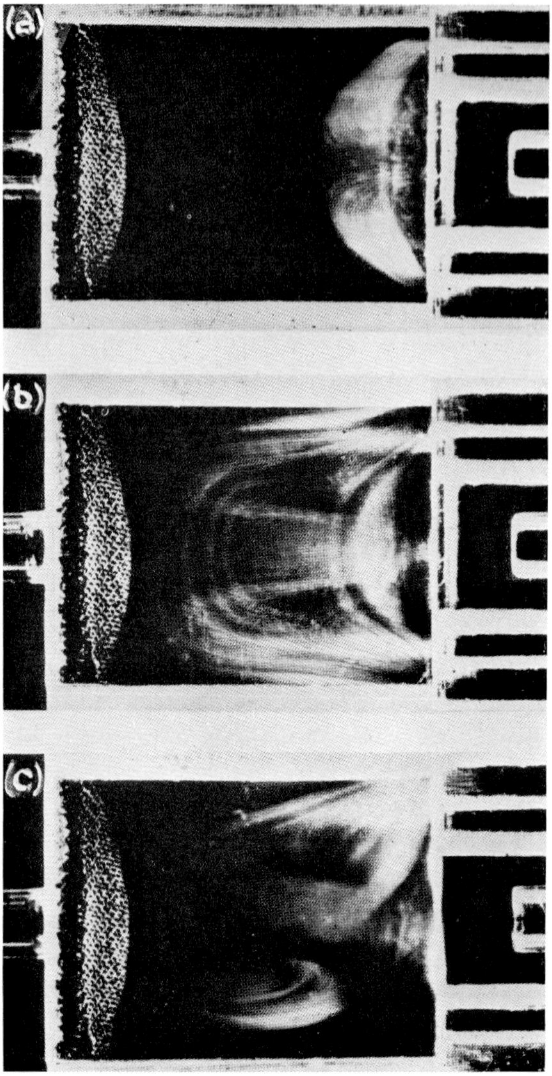

FIG. 2. The effect of detector flow conditions on sample position. (*a*) Anode flow rate: 20 ml./min.; purge flow rate: 100 ml./min. (*b*) Anode flow rate: 20 ml./min.; purge flow rate: 50 ml./min. (*c*) Anode flow rate: 20 ml./min.; purge flow rate: 120 ml./min.

[*To face page 61*

occurs, a profound reduction in the detector current would result when a ^{2}H source is used. The effect on the harder radiation of ^{90}Sr would be insignificant.

In order to establish whether the sample contacts the energizing source, flow patterns within a transparent model of the detector were examined using a smoke visualization technique. The cell was constructed in Perspex and was identical internally with the detector. One side was cut away and replaced by a microscope slide to ensure that the smoke flow pattern was viewed with minimum distortion. Dioctyl phthalate was used to trace the flow pattern and was produced from a monodisperse aerosol generator adjusted to give particles 1 μ in diameter.

Fig. 2*a* shows that at a sample input rate of 20 ml./min. and a purge rate of 100 ml./min., the cloud does not contact the cathode surface where the radioactive source would normally be situated. Fig. 2*b* shows the effect of reducing the purge rate to a value at which the interior of the detector is almost filled with smoke and Fig. 2*c* shows the turbulent conditions resulting from too high a purge flow.

Correlation between the electrical and optical experiments suggests that negative signals arise from adsorption on the tritium source, but when this is prevented positive signals are given. A similar effect might be expected with other strongly adsorbing vapours and sources emitting soft radiation.

SUMMARY

Experiments were undertaken to investigate the effect of water vapour on the performance of a micro-argon ionization detector. The flow field within a transparent model was studied using a smoke visualization technique, and the effects of varying the flow conditions correlated with the observed signal.

Results indicate that correct operation is achieved only when the flow rates are such that the sample does not come into contact with the radioactive source. In these circumstances, the signal is positive with reference to the standing current. Negative responses result from incorrect flow conditions.

REFERENCES

HILL, D. W., and NEWELL, H. A. (1965). *Nature, Lond.*, **205,** 593.

HURST, G. S., BORTNER, T. E., and GLICK, R. E. (1965). *J. chem. Phys.*, **41,** 713–719.

LOVELOCK, J. E. (1958). *J. Chromat.*, **1,** 35–46.

LOVELOCK, J. E. (1960). In *Gas Chromatography*, pp. 16–29, ed. Scott, R. P. W. London: Butterworth.

LOVELOCK, J. E. (1961). *Analyt. Chem.*, **33,** 162–178.

MACKAY, D. A. M. (1960). In *Gas Chromatography*, pp. 357–359, ed. Scott, R. P. W. London: Butterworth.

SANDERS, F. W., HURST, G. S., and BONNER, T. E. (1960). *Chem. Abstr.*, **54**, 2005c.
SHAHIN, M. M., and LIPSKY, S. R. (1963). *Analyt. Chem.*, **35**, 1562.
STRICKLER, T. D., and ARAKAWA, E. T. (1964). *J. chem. Phys.*, **41**, 1783–1789.
SWOBODA, P. A. T. (1960). *Chemy Ind.*, 1262–1263.
WILKINSON, D. H. (1950). In *Ionization Chambers and Counters*, p. 40. London: Cambridge University Press.

DISCUSSION

Hill: I must congratulate you on this technique; I would like to have a smoke generator in our laboratory. In our attempts to understand the working of these detectors we have to take into account a combination of factors, including flow patterns and field distribution. Some of the most bizarre effects occur with this detector with varying purge flow. The length of the purge-flow pipe, the depth of the anode recess and the purge flow itself are all important. It is thus particularly useful to be able to see what is happening inside the detector.

Purnell: But why do you need such a complicated detector at all?

Warren: We chose the micro-argon detector because of its catholic properties and quite high sensitivity. Other detectors are unsuitable for various reasons. The flame ionization detector is unsuitable because its response varies with the degree of substitution in the test compound. Similar considerations apply to the use of the electron capture detector, this being unusable because of differences in electron affinity of the test substances. In some connexions it is advantageous to use a detector whose response is independent of chemical structure and molecular weight.

Purnell: The difficulty is that while your choice of detector is dictated by the problem you are trying to solve, your method is inappropriate because water will inevitably be present in the sample. Water is very efficient at degrading the energy of excited species. These two facets of the problem seem irreconcilable.

Warren: I agree that this is a problem. We have met great difficulties in our attempts to remove water from the system. We are currently trying to chromatograph water, and have found considerable hold-up of water on what one would normally regard as unreactive surfaces. Using a polytetrafluoroethylene cross-over valve, for example, we found that if one line was left to come to equilibrium with water vapour and then switched to the detector, the desorption could be followed; whereas if water vapour was instantaneously injected so that it did not have time to reach equilibrium with the walls of the valve, then good sharp chromatograms were obtained. We have used polytetrafluoroethylene columns, but even with these it is difficult to remove water quickly, that is, in a matter of seconds.

Purnell: Did you imply that the presence of either consistently positive or consistently negative signals does not surprise you but that a mixture of the two does?

Warren: Not exactly. We wanted to determine the exact effects of water in the detector. We knew that water could quench the signals, but we wanted to know also if it could produce a signal in its own right which would have to be added or subtracted from the figures obtained.

Lipsky: Many pitfalls have to be avoided in this system; and many parameters have to be considered. Any alteration in anode design or shape or positioning, for example, will have a profound effect on responses.

Warren: We have not studied the effect of detector material on the lifetime of argon metastable states because we thought that the different signals produced by water vapour could be explained by adsorption on the radioactive foil. With the micro-argon detector the foil covers the whole of the cathode. In later experiments, with a different design of detector, we set the foil back behind the sample inlet to prevent the sample from contacting the source. I am not suggesting that this particular design is not subject to the effects you have mentioned, Professor Lipsky, only that the flow conditions within the detector are more easily controlled.

Scott: I can understand that you have designed a detector and adjusted the flow so that the water vapour could not reach the radioactive foil and thus, in effect, reduced the negative side of the peak. But with carbon-containing material, or an alcohol peak studied first with and then without water in the system, what is the net effect of the water? Is there a significant difference in the response to, say, benzene or ethanol when water is present in (*a*) the old type and (*b*) the new type of detector?

Warren: There is a difference. We have not studied the effect of quenching in any detail although I can say that it may be less than we would expect. If water and organic vapour are admitted together and water gets on to the foil then the organic material may be missed. The difference between the two detectors depends on the lack of quenching effect. If water quenches, one would expect it to remove electrons that are otherwise available for ionization of the organic material. It is difficult to reconcile this effect with positive signals.

Scott: This aspect of your work needs more study.

Warren: These findings were incidental to our main study and it has been difficult to find time to pursue them further; I hope to be able to do this sometime.

Purnell: The flow is an extra—and rather intangible—variable; it might be useful to repeat these experiments with a static system. It seems an extra complication that you have to put smoke into the detector to get an idea of what is happening inside it. Couldn't you just put the sample in, stop the flow and see how the current looks then?

Scott: Water would inevitably get on to the foil in such conditions.

Purnell: Yes, but the presence of water on the foil may not be the root of the problem. The difficulties may be caused by some complex errors in the machine or associated with changes of gas density in the system.

Lipsky: Or these problems may depend on the chemical state of the sample at a particular instant. A number of complex reactions are occurring simultaneously in your experimental set-up.

Purnell: A system that might be useful is one in which flow kinetics are kept constant by the use of a jet down the middle of the tube, while the diameter of the jet relative to the distance to the wall is varied.

Warren: Such an experiment would be valuable although, when the internal diameter of the sample inlet is reduced, negative regions are reached sooner. The smoke in such an experiment would have a higher linear velocity than in the experiment I described, wouldn't it?

Purnell: That is not the right experiment because the diffusion path would be maintained at about the same length, which means that you would be sampling closer to the turbulent region in the stream. What is needed is an experiment carried out at the same jet size but with varied distance to the wall so that the length of the diffusion path is increased. This would tell one unequivocally if diffusion of water to the wall is important or not.

Scott: The experiments Mr. Warren carried out show fairly conclusively that there is a significant difference in the response of the detector when water is kept physically away from the source. It is surely also significant that a water column always gives positive signals in these experiments.

Purnell: I agree.

Warren: When we used the micro-argon detectors with packed columns and dispensed with the purge flow, that is, we admitted the sample through the inlet, we found that water vapour produced only negative signals.

Purnell: The converse of that might be that all the negative signals arise when water is removed from the wall or gets to the foil. The amount of water the foil can absorb is limited—one cannot build up an indefinite number of multilayers. Eventually, in continuous operations, one would expect a stable state to be reached.

Warren: But at that limit one might have stopped the electrons emanating from the foil.

Purnell: I had not anticipated that.

Scott: This would happen with tritium but not with strontium.

THE W-VALUE DETECTOR. DETERMINATION OF OXYGEN AND ANAESTHETIC VAPOURS IN EXPIRED AIR

B. C. H. WARREN AND J. E. LOVELOCK

Physics Research Division, Chemical Defence Experimental Station, Porton Down, Wiltshire

PERHAPS the simplest of all ionization phenomena, in gases, which can be exploited for the detection of one or more components of a mixture is the W-value. This is the energy absorbed from an ionizing particle traversing the gas during the production of an ion pair; it is a characteristic property and varies from about 25 eV for organic vapours to over 40 eV for pure helium. A detector exploiting such a property possesses two features which make it unusually valuable in certain applications. First, it is completely insensitive to large changes in temperature, pressure and gas flow; and second, its response to each of the common gases and vapours of the atmosphere—nitrogen, carbon dioxide and water vapour—is similar so that variations in their concentration do not affect the detection of other gases and vapours to which it responds, for example oxygen and organic vapours.

Detectors

A prototype of a W-value detector was described by Lovelock (1965), but any ionization chamber may be used provided certain conditions are observed. These are: (*a*) the chamber dimensions must be chosen so that, under the operating conditions contemplated, the range of the ionizing particles is contained by the chamber; (*b*) the electrical field must ensure the collection of all the ions formed without loss by recombination or gain by ion multiplication; and (*c*) the source of radiation used must be constant in energy and intensity. If these requirements are satisfied the current flow with an appropriate polarizing potential will be reciprocally related to the W-value of the gas.

Operating potential is determined by the nature of the atmosphere to be sampled and by the dimensions of the detector. In general, a potential sufficient to collect all the ions produced by the radiation at the highest pressure and highest gas concentration is chosen. This ranges from about 30 V for a 3-mm. electrode separation to 700 V for a 50-mm. separation.

Fig. 1 shows the detector used in these experiments. When filled with nitrogen and operated at 60 V, a 200-mCi tritium source produced a background current of $1 \cdot 4 \times 10^{-8}$ A.

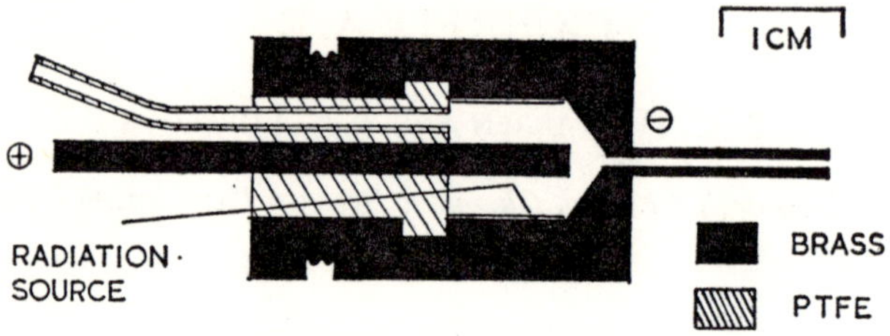

FIG. 1. W-value detector.

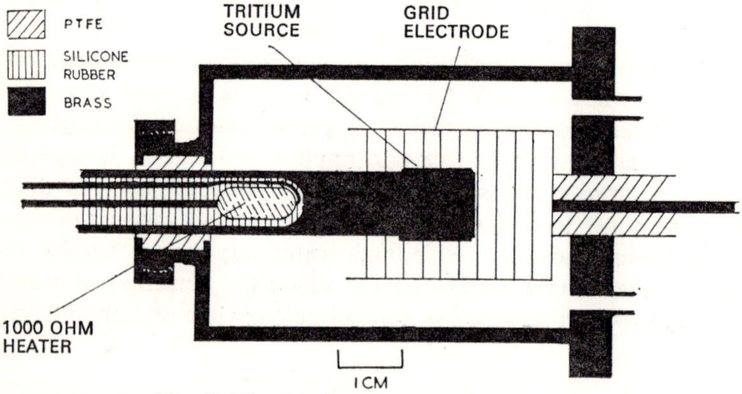

FIG. 2. W-value detector—practical version.

An alternative form of detector closely resembling Lovelock's prototype but incorporating a small heater to prevent condensation of moisture within the detector is shown in Fig. 2. Condensation on the radiation source may cause a false negative signal and on the detector insulation a false positive. The favourable position of the source in this arrangement ensures that cross-sectional effects are absent. In certain circumstances it may be more convenient to use an open form of the detector which is continuously exposed to the test atmosphere, thus obviating the need for a sampling pump.

Performance

The indifference of the detector to changes in temperature and pressure and its response to various substances are shown in Table I.

TABLE I

CHARACTERISTICS OF THE W-VALUE DETECTOR

Background current in air	$1 \cdot 4 \times 10^{-8}$ A
Noise level at a time constant of 2 sec.	10^{-12} A
Temperature and pressure range for < 0·1% change in detector current	
Temperature (°C)	20–200
Pressure (mm. water)	60–760
Test Substance	*Increase in ion current when 10% concentration of test substance in air is added (per cent)*
Oxygen	0·7
Water	0·04
Carbon dioxide	0·15
Methane	3·2
Propane	5·0
Heptane	8·0

The response of the detector is linear to 100 per cent of any detected compound as illustrated for oxygen and cyclopropane in Figs. 3 and 4. Although the signals are small (2×10^{-10} A for a 20 per cent concentration of oxygen and 7×10^{-11} A for a 1 per cent concentration of a typical anaesthetic vapour) the insensitivity of the detector to changes in the ambient variables makes their measurement possible with good accuracy and neglible drift.

Response factors for the various anaesthetics differ over a range of approximately two and are as follows: chloroform 3; diethyl ether 6·5; cyclopropane 5; halothane 4; methoxyflurane 4·5.

The unique characteristics of this detector make it possible to construct a simple apparatus for monitoring continuously the concentration of oxygen

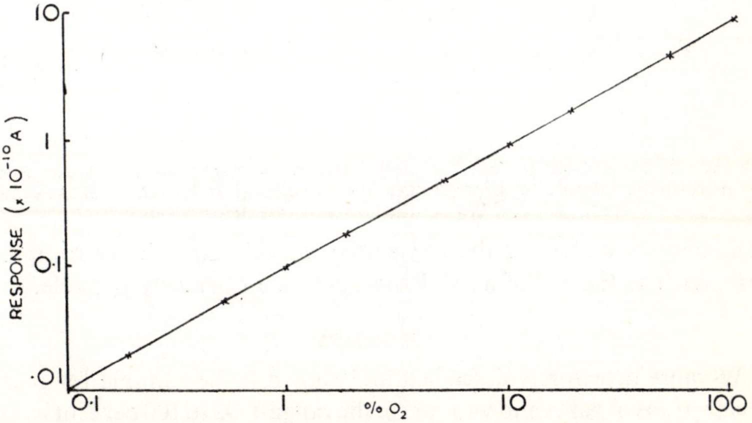

FIG. 3. Response of the W-value detector to oxygen.

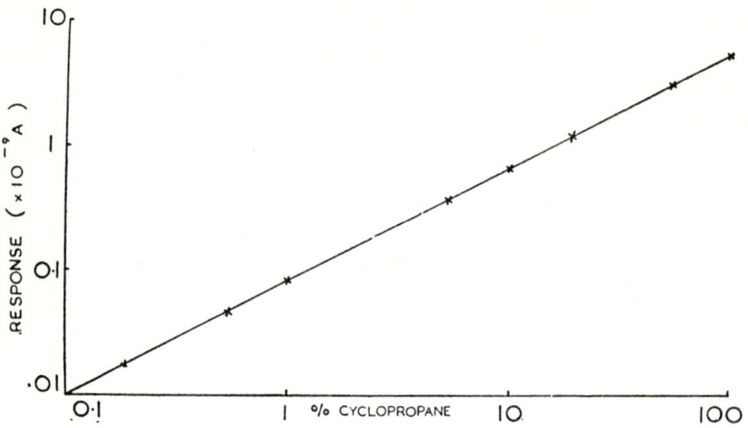

FIG. 4. Response of the W-value detector to cyclopropane.

and anaesthetic vapour in respired air. To achieve this, two W-value detectors are used in series, with a trap between them to remove anaesthetic vapour. Fig. 5 shows a block diagram of such an apparatus together with the electronic arrangement needed to provide a continuous indication as a meter reading of both oxygen and anaesthetic vapour. In the design shown the anaesthetic

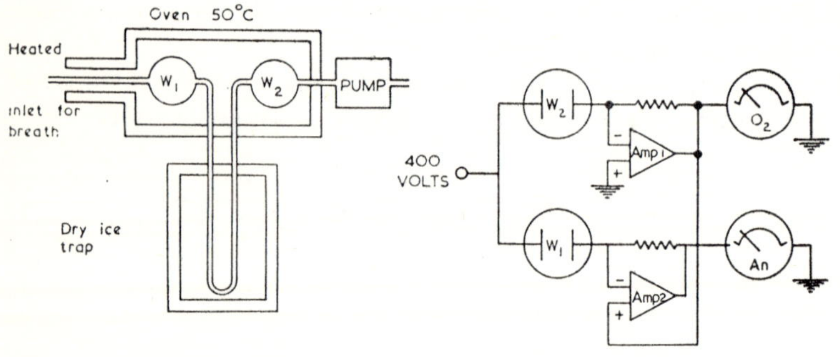

FIG. 5. Composite system for simultaneous measurement of oxygen and anaesthetic vapour. (*a*) Block diagram; (*b*) electronic arrangement of the same system. W_1 and W_2 are the two W-value detectors in series.

vapour is separated from the oxygen by a cold trap. Other trapping procedures such as the use of a solid adsorbent are obviously possible.

SUMMARY

A W-value detector is described and shown to be suitable for measuring concentrations of gases and vapours in the range 0·02 to 100 per cent by volume, the minimum detectable oxygen concentration being about 0·1 per cent.

The detector is unaffected by large changes in temperature, pressure and gas motion and is insensitive to variations in the proportions of the common gases and vapours of the atmosphere, including carbon dioxide and water vapour.

The high stability of the detector enables a simple apparatus to be made for monitoring continuously the concentration of oxygen and anaesthetic vapour in respired air, thus eliminating the need for a chromatographic column.

REFERENCES

LOVELOCK, J. E. (1965). *Analyt. Chem.*, **37**, 583–584.

DISCUSSION

Hill: How do you arrange this system so that one detector responds to cyclopropane and the other to oxygen?

Warren: They both respond, but in the first detector the combined current from both gases and in the second only the current from the oxygen are measured.

Hill: Could this system be made more selective?

Warren: No. The detector in the form I described is calibrated for either halothane or cyclopropane; one cannot discriminate between the two. But one could use a number ($n - 1$) of detectors and thus sample other materials although one would need an external separating procedure to do this.

Mitchell: Have you any figures on the quantitative reproducibility of this system?

Warren: No; I have only been using it for three weeks, but in this short time our results have been consistent. With the open form of detector (in which diffusion is used) the standing current gives rise to the same signal, quantitatively, for up to three hours. This is a stable system. Similarly, changing the dimensions to eliminate pressure effects does not alter the performance.

Payne: Have you used this system in man? The response during spontaneous breathing would be interesting.

Warren: This is not my province. We were, of course, primarily interested in other forms of detectors. We have used this device for hydrocarbons and a few anaesthetics but have not tried it in the human subject.

Lowe: Does it respond to nitrous oxide?

Warren: A rough guide is that the W-value is related to the ionization potential. So this system responds to nitrous oxide but less sensitively than it does to halothane and cyclopropane.

Hill: What sort of amplifier are you using? This is clearly a small robust detector, and, as such, needs a small robust amplifier.

Warren: We use a small operational amplifier (Analog Devices, model 310), with an offset current of about 10^{-14} A, costing less than £30.

Lipsky: Many governments are looking for a portable detector that could be left in the field to monitor trace quantities of toxic compounds. But the W-value detector could also be useful to the practising anaesthetist. Despite modern advances, anaesthesiology is old-fashioned in many ways and a precise method for detecting the concentrations of expired anaesthetics would be valuable.

Payne: A method for obtaining concentrations virtually on-line is urgently needed in anaesthesiology. The difficulty with present instrumentation is that the processing of samples is still too slow to benefit individual patients. Thus, although it is possible to derive adequate statistical analyses for long-term interpretation, we are not yet in a position to use this information in the minute-to-minute management of patients.

Warren: This is the advantage of the W-value-detector arrangement I have described: the continuous meter reading follows almost immediately on any change in concentration.

Scott: The breath analyser also gives a quick response.

Payne: Could the W-value instrument be used to detect alcohol in the breath? If so, this might solve quite a few problems.

Warren: Yes. Although we have not analysed alcohol in the breath with this device, it is sensitive to ethyl alcohol vapour, and the other normal components of expired air would not interfere with the detection of alcohol. But the W-value detector could not be used simultaneously for ethyl alcohol and anaesthetic gases without a further separating stage.

A. Curry: The great problem in the determination of alcohol in the breath using cheap detector tubes is that the method involves blowing up a 1-litre bag; thus the sample contains a variable proportion of dead-space air. Even with a tube that gives a perfect reading there is no way of knowing what proportion of the total is alveolar air. A device that can measure, instantaneously, the alcohol level exactly at the alveolar plateau would be an exciting new development.†

Payne: One would have to make the assumption that the partition coefficients are correct.

A. Curry: Arbitrary baseline figures would be needed, of course.

Robinson: We have just completed an experiment which illustrates the unreliability of measurements taken from the breath analyser. (Incidentally, this information may provide a useful way of "beating the breathalyser"!) The lung itself can be visualized in three zones. In the upright lung, the lower

† See pp. 80–81 for further comments on the alveolar plateau.

zone is always perfused with blood, the middle zone occasionally, and the upper zone not at all. When an individual inspires a single normal breath air reaches the middle and lower zones only, but if he breathes out until only the residual volume of air remains in the lungs and then takes in one litre of air the situation is rather different: nearly all this inspired air passes to the upper zone because of compliance changes in the lung at low lung volumes, but the upper zone is unperfused and alcohol cannot, therefore, be transferred from blood to breath. We have checked the alcohol concentrations which would be measured by the breath analyser with a mass spectrometer and have confirmed that, after a maximum expiration to reduce lung volume in this way, alcohol levels as measured by the breath analyser are deceptively low. In practice, if, after alcohol, one breathes out deeply, then inspires about one litre and immediately blows into the breathalyser bag, the alcohol concentration in the bag may not show an increase.

Hill: Some years ago (Hill, Hook and Mable, 1962) we analysed gases very rapidly using a 2-metre open tubular column with a cathode ray tube display and a micro-argon detector. We could accurately measure halothane vapour concentrations in the laboratory with this instrument. But when we tried to use it in a clinical set-up, we recorded values of around 3 per cent for the concentration of halothane when the patient was in fact breathing about 1·5 per cent of this vapour. This inaccurancy was due to water vapour, but we also had unexpected anomalous effects with nitrous oxide and similar gases (Hill and Newell, 1967). This has been the trouble with many of our instruments when applied in the clinical field. We need a small portable device that uses a non-explosive gas. Although metal oxide, silicon, transistor amplifiers are available we still cannot use most of these instruments without a flame, which is not allowed in the operating theatre. And one meets so many unexpected problems; I would never have suspected trouble from nitrous oxide. Nor can we specify the condition of the patient. We cannot, for example, stop him from breathing out *moist* air. Although a general-purpose chromatograph with a flame detector attached is adequate for most laboratory tests, problems arise when we try to devise a suitable instrument for clinical studies.

Payne: Basic scientists do not always appreciate that clinical measurements are still fairly crude. Inspired concentrations of halothane, for example, are of the order of 1–5 per cent, but with a margin of error of ± 1 per cent.

Purnell: Dr. Hill, would you amplify your comments about your difficulties with nitrous oxide?

Hill: The sensitivity of the detector to halothane and ether increased in the presence of nitrous oxide (and also, incidentally, of carbon dioxide).

Purnell: Was this because you did not resolve the several components or because of something more mysterious?

Hill: The increase in sensitivity is marked at low field strengths; this is probably because of an increase in the effective free electron paths. A small negative peak for air occurred with all the columns we used. The permanent gases (oxygen, nitrous oxide and carbon dioxide) eluted first and were followed by the organic compounds (cyclopropane, ether and halothane). I assumed that the permanent gases had passed completely through the detector whereas, in fact, some remained in it.

Purnell: This could be because of the high solubility of nitrous oxide in organic solvents; it is quite difficult, for example, to separate nitrous oxide from ethylene with a squalane column.

Hill: Another possible explanation might be connected with sample retention by the detector. Or we may have to look for something more subtle. In some experiments with sampling during anaesthesia I assumed that by the time the initial small gas peak appeared the bulk of the gases had passed through the column; but with later static experiments using an ionization detector some bizarre effects that were dependent on the polarizing voltage were shown. These findings suggest that an increase in the peak for the anaesthetic vapour might be associated with low polarizing voltages. The explanation for these contradictory observations is probably that we had increased the effective free path of the electrons at low field strength. But what is so interesting is that even though the column had separated the gases from the organic materials, they (the gases) were still present.

Scott: Nitrous oxide is retained for a long time in the system because it is so soluble in inorganic solvents. In its preparation, for example, it must be collected over hot water because it dissolves in cold water. In Mr. Warren's experiment the nitrous oxide is probably not eluted from the liquid column until later on in the chromatogram because it is so soluble.

Another interesting point you raised, Dr. Hill, is the taboo on the use of flame ionization detectors in the operating theatre, presumably because of fear of fire. Yet the presence of the flame ionization detector in petroleum refineries, where the biggest fire risk possible exists, is now accepted as perfectly safe. Why is the flame ionization detector considered to be safe in petroleum refineries but not in an operating theatre?

Hill: Flame-proof detectors are available but there is still the risk of a hydrogen leak from a loose coupling. On balance, it seems safest not to use hydrogen in the operating theatre.

Scott: If one used an electrolytic system for producing hydrogen and oxygen, and apparatus is available that can produce adequate amounts of these gases, one would not need a gas cylinder at all.

Warren: In the W-value system we need no cylinder at all because we sample direct from the atmosphere.

Lipsky: A long tradition, derived from the days of open ether anaesthesia, lies behind the fear of fire and explosions in the operating theatre.

Payne: This is a medicolegal hazard. In the UK the use of explosive gases in the operating theatre is rigidly controlled by Ministry of Health regulations.

Scott: But is there now a scientific reason for this taboo?

Payne: No.

Purnell: In 1961 it was axiomatic that no flame detector would ever get inside any chemical plant in the US, simply because insurance companies would not insure for this risk. And taboos involving money are notoriously difficult to break down! But in 1969 all chemical plants have a flame detector, which now turns out to be an explosion-free, hazard-free and generally excellent device. It is defeatist to say that because hydrogen has been barred from operating theatres in the past it must continue to be thus prohibited.

I suggest that in the practical use of any of these techniques in the operating theatre, the inspiration for development is most likely to come from the field of process instrumentation rather than laboratory instrumentation, because so many of the relevant clinical problems are similar to problems in industry. For example, the patient expires wet air—most petroleum refineries handle wet petroleum; and industrial chromatographic instruments must be able to meet this situation or they will not sell. The existing expertise in the processing side of chromatography could probably be profitably applied to the clinical field.

Payne: The availability of funds within the National Health Service in the UK is also relevant to this problem. Money for research is limited and, perhaps not unnaturally, what there is tends to be spent on clinical research rather than on instrument development.

REFERENCES

Hill, D. W., Hook, J. R., and Mable, S. E. R. (1962). *J. scient. Instrum.*, **39**, 214–216.

Hill, D. W., and Newell, H. A. (1967). *J. Chromat.*, **32**, 737–739.

THE GAS CHROMATOGRAPHIC COLUMN AS AN ANALOGUE FOR RESPIRATORY FUNCTION IN THE LUNG IN MAN

J. JANÁK

Institute of Instrumental Analytical Chemistry,
Czechoslovak Academy of Sciences, Brno

WHAT useful information can we derive from considering the gas chromatographic column as an analogue for respiratory function in man? The use of gas chromatography in the analysis of gases in blood and breath is well known, and described in a series of papers in the literature on physiology. I do not intend to give a detailed description of this application of the technique but would like to emphasize the potential use of gas chromatography as a method not of analysis of pulmonary gases, but rather for study of the respiratory process in the lungs.

Let us first abandon our idea of gas chromatography as an analytical tool and study the possible relationship between the physical pattern of the gas chromatographic column and the lungs, looking for analogies between the column and lung tissue and deriving new analytical approaches to the study of the respiratory process (*a*) in healthy and damaged lung tissue, and (*b*) in the effusion of volatile constituents from the pulmonary tissue of healthy and sick human beings.

The gas chromatographic column is characterized by a liquid, coating a large area of porous support. The liquid forms a thin film allowing the establishment of partition equilibrium of the solute between the carrier gas and the liquid. From the retention of a component we may derive the extent of partition of the solute between the two heterogeneous phases. The shape of the peak permits the acquisition of knowledge about the sorption isotherm and the rate of establishment of the equilibrium, and also about the resistance to mass transfer.

To what extent does this model approach an expression of the function of lung tissue? The area of lung tissue is extremely large. The lung, also, is a heterogeneous system whose interface is composed of blood corpuscles and gas and the gas can be either air or an appropriate gaseous mixture.

The blood plays the role of solvent in this system. It sorbs, physically or chemically, oxygen, carbon dioxide, carbon monoxide, inert gases and other gases (*vide infra*). A change in the composition of the inspired air will occur if the air remains in the lung for an appropriate time. Here an obvious analogy exists in the establishment of sorption equilibrium in a gas chromatographic column and in the lung.

As the sorption equilibrium given by the partition coefficient in a gas chromatographic system reflects both the properties of the solvent-solute system and also, at a known concentration of solute, the quantity of solvent, a change in the composition of inspired air in the lung may similarly reflect the quality and quantity not only of haemoglobin but also of other blood constituents of the pulmonary tissues.

Gas chromatographic examination of pulmonary action

Let us now consider a situation in which time is insufficient for, say, the establishment of the partition equilibrium. The time profile of the sorption of gaseous constituents by blood in the lung tissue might give information about the properties of the interface, that is, about the health and integrity of the lung tissue. Here we see an analogy with the derivation of the term "resistance-to-mass-transfer" from the width and shape of the gas chromatographic peak.

In the first situation, detailed analysis of the total expired gases is generally employed to determine gas exchange between air and blood. In the second, the composition of the expired gas at various moments of the respiratory cycle must be determined. A gas chromatograph for this purpose is shown diagrammatically in Fig. 1. It is designed for both manual and automatic withdrawal of a number of samples at pre-set time intervals. The patient breathes either air or an artificially made-up mixture of gases, and the expired gases are sampled. The carrier gas is bypassed from the inlet to the appropriate chromatographic column system and thence, through a katharometer, into the atmosphere. The diaphragm valves for sampling one to nine steps each respiratory cycle each have a volume of 0·4 ml., are controlled electronically and can be opened separately one at a time, manually or automatically, at pre-set intervals from 0·1 sec. to about 9 sec. The expired air flows through the sampling bypass and is trapped in the sample loops. The special injection-port for manual introduction serves for the calibration of the chromatograph. The chromatographic columns are tubes, 2 mm. in diameter, packed with Porapak Q, molecular sieve 5A and, for example, dioctyl phthalate. A diagram of breath sampling is shown in Fig. 2. *A* illustrates the feasibility of obtaining up to nine samples during an expiration lasting

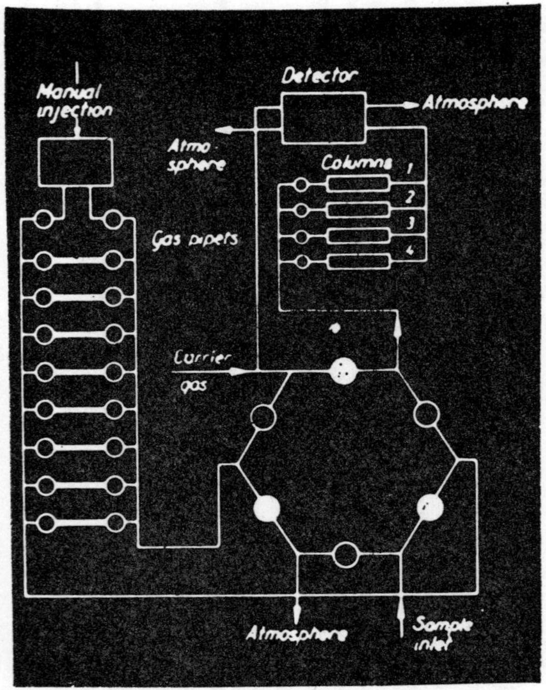

FIG. 1. Scheme of a gas chromatograph for the investigation of pulmonary action. Left: injection-port and nine diaphragm sampling valves; right: stopcock manifold; top right: column system and katharometer.

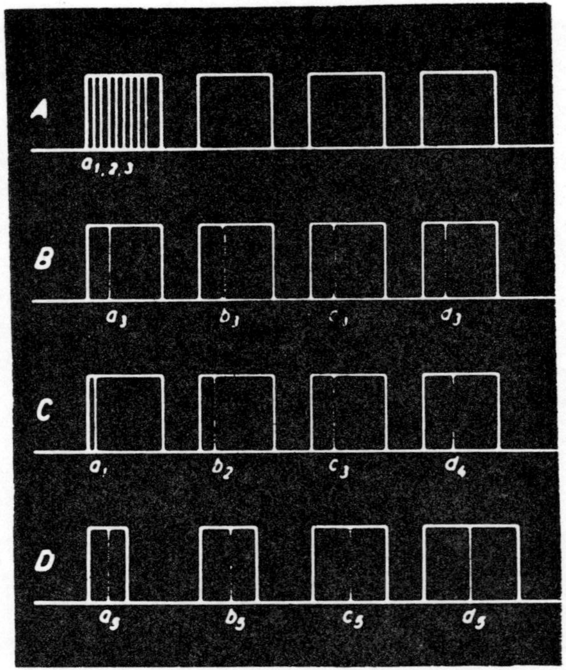

FIG. 2. Schematic representation of methods for sample withdrawals. (For explanation, see text).

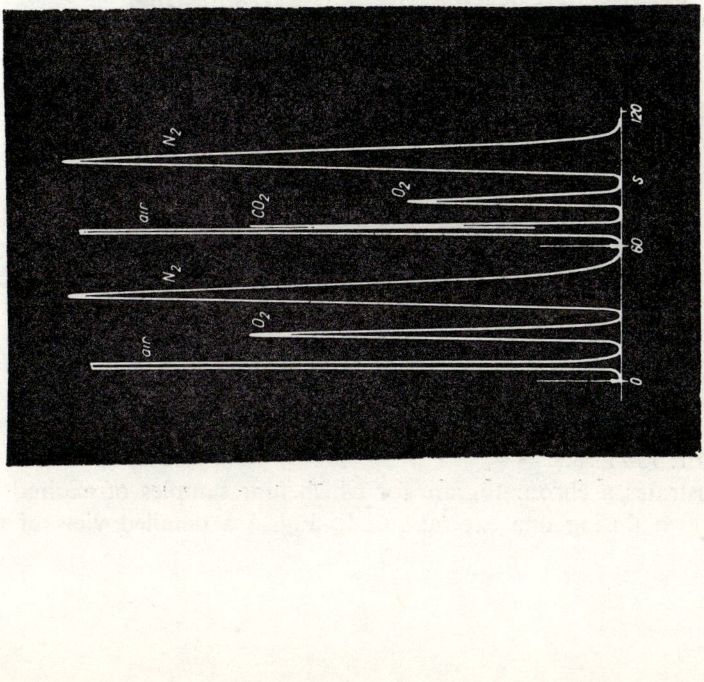

FIG. 3. Chromatograms of inspired air (left) and one sample from expired air (right) developed on two parallel working columns (Porapak Q and molecular sieve 5A).

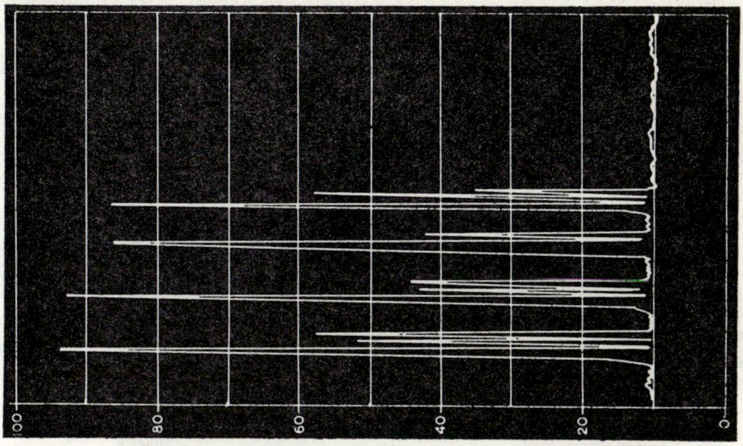

FIG. 4. Chromatograms of four samples withdrawn from one expiration.

1–2 sec.; *B* shows the withdrawal of single samples from the same stage of consecutive expiratory cycles; *C* shows withdrawal of only one sample during each expiration, taken at various stages of the expiratory phase and *D* shows the withdrawal of single samples from the same stage of consecutive expirations after various intervals of breath-holding.

Fig. 3 shows chromatograms of atmospheric air, and one sample of expired air, separated on two parallel working columns packed with Porapak

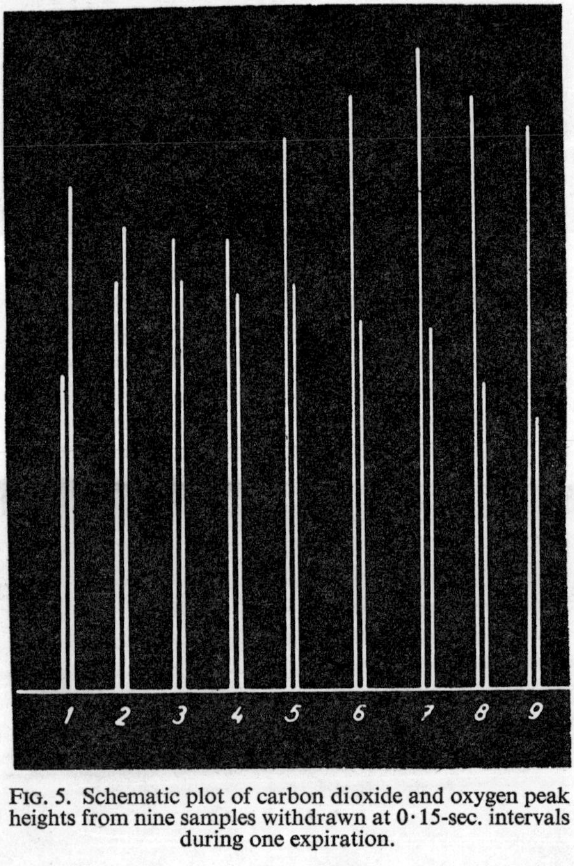

FIG. 5. Schematic plot of carbon dioxide and oxygen peak heights from nine samples withdrawn at 0·15-sec. intervals during one expiration.

Q (diameter: 2 mm.; length: 250 mm.) and molecular sieve 5A (diameter: 2 mm.; length: 750 mm.).

Fig. 4 illustrates a chromatogram for which four samples of expired air were withdrawn during one expiration. In Fig. 5 a detailed view of the composition pattern of expired air during one expiration is shown. The individual chromatograms represent the composition of the expired air at consecutive moments, 0·15 sec. apart.

Applications

I would now like to mention further potentialities for the analytical model I have described. As this experimental set-up permits examination of pulmonary function at very short exposures, one can use quite uncommon gaseous mixtures to study the quality and function of pulmonary tissue. The presence of an appropriate test gas, for example a hydrocarbon such as butane, in the inspired air offers an interesting means of characterizing the type and extent of a pneumoconiosis, generated by coal dust or silicate and other dusts. Narcotics, or other toxic compounds such as acetylene, cyclopropane, nitrous oxide or carbon monoxide might also be used in particular cases.

I shall conclude my presentation by mentioning another aspect of the analogy between the lungs and the gas chromatographic column. The vapour pressure of solute over its solution is determined by the properties of the system and the concentration of solute in the solution. The concentration of vapours over blood will change similarly depending on their concentration in blood. Therefore, if the blood of an ill patient contains abnormal volatile constituents, and normal volatile constituents in abnormal amounts, the established equilibrium between blood and pulmonary gases must somehow reflect this. Positive qualitative results in patients with diabetes or hepatic coma may confirm this hypothesis. Quantitative interpretation is difficult because blood is a colloidal solution containing solids of varying, but always high, molecular weights in an aqueous solution.

DISCUSSION

Hill: Will you expand your comments on the sampling valve? If you took nine samples from one breath you would presumably have to store these temporarily, before they were fed into the chromatograph, because the analysis time for the chromatograph would be longer than the time taken for each breath. The key to this difficulty must be the sampling valve.

Janák: You are right. The samples are stored until the analysis can be carried out, although in practice this can be done almost immediately. The valves are membranous and electronically operated, so they can be opened at very short intervals. A small vacuum is needed on the distal end of the gas-pipette line so that rapid filling of the tubes (during 0·1 sec.) can take place. The pressure of the breath is not sufficient to fill the tubes.

Payne: This technique raises all sorts of interesting possibilities; it might replace the original method of using acetylene for calculating cardiac output (Grollman, 1929). It should be much easier to determine the uptake of acetylene from the lungs with this sort of technique than with Grollman's method.

Lipsky: Many previous attempts to automate this sort of technique for clinical use have been unsuccessful. I know of no centre of anaesthesiology in the US where this technique is used routinely for monitoring in patients.

Payne: We have to deal with two basic problems: first, many clinicians are suspicious of the value of measurement techniques and, second, during elective surgery the preparations for this type of study often take longer than the operation itself. Surgeons are usually unwilling to put up with the associated inevitable delays unless they themselves have had some training in investigative methods.

Lowe: These problems are the responsibility of the anaesthetist who elects to do this type of monitoring. It is a routine procedure in our operating theatre, where we use both a chromatograph and a hydrogen flame detector. We take samples of gas into our instruments every three or four seconds and analyse blood intermittently every few minutes. This schedule is designed not to interfere with the surgeon's routine and can be efficiently done without difficulty.

Payne: This is easier said than done in the UK where, on the whole, operating time is shorter than in the US. This may be due to the different ways in which surgery has developed in the two countries. Moreover, the fact that one surgeon commonly completes the whole operating list on this side of the Atlantic tends to shorten the intervals between cases and further limits the anaesthetist's opportunities.

Hill: In Professor Lowe's department I was once able to record chromatographically the uptake of an anaesthetic during an eight-hour operation. Such an opportunity would never occur in England.

Sjövall: Samples can be taken into the chromatograph in a matter of seconds. This surely would not interfere too much with any operation.

Lowe: No, but the advantage in the US is that because operating is slower we have a longer time for our analyses. With a competent technician there is no question of slowing down the surgeon; indeed, he may gradually come to appreciate the presence of our instruments.

Robinson: What we are really interested in is the alveolar plateau, and the gas samples obtained from breath from the mouth are not derived solely from this plateau. In my department we are now using a four-channel mass spectrometer as a rapid gas analyser. This instrument provides an improved way of assessing the alveolar plateau. A great deal of information is thrown away when intermittent sampling and recording on the gas chromatograph are used.

Scott: What is the alveolar plateau?†

Robinson: It is an expression of the gaseous composition of the alveoli. The lungs and respiratory tract can be divided into three zones: the upper air passages, trachea and large bronchi—the dead space; an intermediate zone occupied by the medium sized bronchi and bronchioles—the zone of linear mass transfer; and, finally, the alveoli themselves where gas exchange with the blood takes place. The composition of gases in the dead space is the same as that of atmospheric air. The zone of linear mass transfer contains a mixture of air from the dead space and the alveoli. The mixture of gases in the alveoli represents the equilibrium with the blood. Random sampling of expired air cannot identify which of these three zones the sample comes from. Rapid sampling, for example by a mass spectrometer throughout expiration, is therefore needed to analyse the precise composition of the alveolar plateau. Our results will be misleading unless we have a reliable method for the *immediate* analysis of the gases in expired air.

Scott: But surely you can pick out a particular part of expiration from the chromatogram if you sample regularly at short intervals and thus follow the respiration pattern.

Robinson: We would still only be guessing. It is conventional to obtain air from the alveolar plateau during the vital capacity manoeuvre (that is, during expiration after maximal inspiration) in the following way: the first 750 ml. of air expired during this manoeuvre comes from the first two zones of the lung (the dead space and zone of linear mass transfer) as described above; the next 500 ml. comes from the diffusion zone; thus, in healthy human beings, samples of air taken between 750 and 1250 ml. are considered to come from the alveoli themselves. But human beings, especially when they have been given an anaesthetic, are not healthy; we cannot select only "normal" patients whose alveolar plateaux are between 750 and 1250 ml.

Scott: Dr. Janák's curves show the entire composition of one breath.

Robinson: But he was still sampling intermittently. We do not even get a plateau with intermittent sampling. The graph for expired oxygen, for example, is a climbing line rather than a level plateau.

Scott: What gases are you interested in?

Robinson: Carbon dioxide, nitrogen and water vapour.

Scott: Under some conditions Dr. Janák's nine samples would each

† See also p. 70.

produce a similar curve for carbon dioxide. One might need to sample from different parts of alternate breaths to give the chromatograms time to develop and thus obtain a composite picture of events during one cycle.

Robinson: The distribution of gases expired from the lung is shown only by the slope and height of the concentration of each gas in the expired sample plotted simultaneously on an X–Y plotter against the volume of gases expired.

Scott: But you can see that already.

Robinson: Yes, but the interval between each sampling is too long; the crux of the problem is that we cannot sample fast enough.

Purnell: Surely this problem depends on the definition of continuous. One complete breath (inspiration + expiration) lasts for about three seconds. A sampling rate of 10 per second gives us 30 single points per respiratory cycle, which is more like continuous monitoring than intermittent analysis. Is the problem that the time scale of the events occurring during respiration is too fast for the gas chromatograph?

Robinson: That is exactly it.

Scott: Does the mass spectrometer record the value for specific peaks or is the instrument scanning continuously?

Robinson: Normally we use four detectors and detect four individual masses. But we can scan with the mass spectrometer if we want to.

Scott: You must also elicit readings of peak heights.

Robinson: Not really; the four detectors are adjusted so that they are sited on four masses and the collectors are mechanically adjusted to peak values for each gas.

Purnell: The cost of obtaining a continuous record of four gases must be at least £40 000 if you use, say, a mass spectrometer such as the MS 12 (AEI model).

Robinson: We use a respiratory mass spectrometer which costs about £5250.

Scott: Such an instrument is much more expensive than a gas chromatograph.

Robinson: Yes, but the results with it are incomparably better.

Scott: Do you get reproducible results on a breathing patient with such small periods of sampling? I would surmise that the curve obtained with a gas chromatograph would not be reliable because the variation in gas composition in the expired air would be far greater than the small variation due to the variability of the instrument itself.

Purnell: To put it another way, is it worth using a really high-class tool—the mass spectrometer—for this rather rough job?

Robinson: I disagree that these analyses are rough. The differences in partial pressures of gases in the blood are (for nitrogen, say,) of the order of 10 mm. Hg.

Purnell: But that is an enormous amount.

Robinson: Not when we consider the solubility of these gases; they are virtually insoluble in tissue fluids.

Purves: In practice, the measurement of respiratory gases involves some compromise with accuracy because of the costs involved. Dr. Janák's method is ingenious but complicated and, as I understood it, offers no obvious advantages over other instruments such as, for example, i.r. or paramagnetic analysers. Further, Dr. Janák's method of describing the profile of alveolar carbon dioxide during expiration is far from being on-line, which is what the anaesthetist requires. Although theoretically the method could be adapted to measure a larger number of samples within each respiratory cycle so that the analysis was virtually continuous and a variety of gases could be analysed, there are simpler ways for getting this kind of information.

Payne: I could accept what you say if these simpler techniques were being introduced. Unfortunately, many clinicians are still unaware of the need for any such techniques or instruments, simple or complex.

Robinson: After a single breath of 100 per cent oxygen we can plot volumes of expired carbon dioxide, rapidly and accurately sampled, against the total volume of expired gases (rather than against time) (see Fig. 1). The measure-

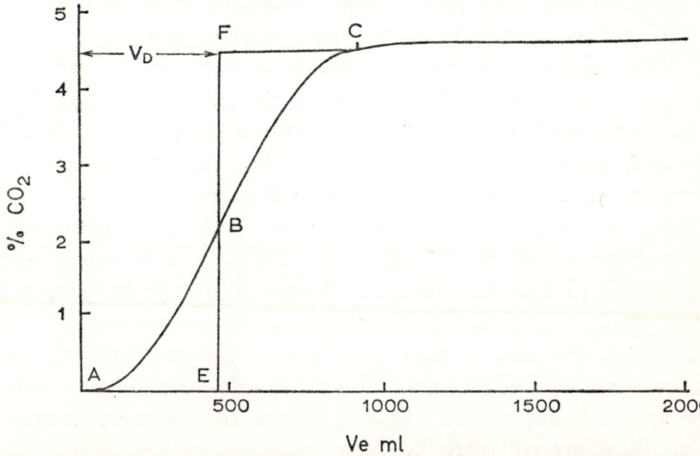

FIG. 1 (*Robinson*). Graph showing volume of expired carbon dioxide (%CO_2) on ordinate plotted against the total volume of expired gases (Ve ml.) on abscissa. Provided area BFC = area ABE, then AE is equivalent to the physiological dead space (V_D). In the case illustrated, with an expired, volume of 2000 ml. V_D = 450 ml. (an extreme example).

ment from *A* to *E* along the abscissa represents the volume of the dead space. This volume is important because if the anaesthetist introduces the anaesthetic and other gases too forcibly, the patient's respiratory efforts are paralysed and the lung may be partly emptied of blood (from the pulmonary circulation). Thus a part of the lung previously available for gaseous exchange now becomes, in practice, dead space. The total volume of dead space is represented by area *ABE* in Fig. 1 and can be measured in millilitres. This volume indicates whether the anaesthetist is being too heavy-handed.

Scott: How much would the dead space have to increase before the anaesthetist realized something was wrong and changed his tactics?

Robinson: The volume of dead space, normally about 150 ml., increases to about 185 ml. during anaesthesia, even in competent hands, and it may be as high as 350 ml., by which time the patient could be dead. But *changes* in the volume of the dead space are too small (of the order of 50–60 ml.) to be detected by a gas chromatograph and we have only recently been able to monitor them. Another way of looking at this is to say that if the dead space becomes too large some of the gases given to the patient are wasted because dead space has no contact with blood and diffusion cannot occur. Fig. 1 shows a hypothetical graph of a single breath analysis of the dead space. The length of the line AE and the climb of the curve are both important. But we need to know and measure the content and partial pressures of gases in the blood as well as in the lung before we can draw any helpful conclusions.

Payne: Incidentally, your department, Professor Robinson, is probably the only unit in this country doing this very specialized work.

Purnell: How can you watch four traces simultaneously and make the necessary deductions about and responses to these subtle changes?

Robinson: We soon become used to looking at the records, which can be picked up quite quickly with a fast u.v. recorder. But it is easiest to use an analogue computer to compute the derived values we need.

Scott: Information produced as quickly as this must be processed automatically because no individual could respond to it sufficiently rapidly. It follows, therefore, that automatic controls must eventually be built into every anaesthetic system.

Robinson: Yes, but this is only one half of the problem; we are only looking at one side of a membrane. The other side—the blood—raises more important and difficult issues. Our methods for measuring respiratory gases in the blood are still appallingly crude.

Cervenko: Mr. Warren's and Dr. Janák's papers suggest new methods which may be more effective than the older techniques of u.v., i.r. or polaragraphic analyses of respired gases. One drawback of these older methods is that one gas may interfere with the detection of another. Carbon dioxide, for example, interferes with certain inhalation anaesthetics on the i.r. analyser. These two new techniques should improve our analysis of respired gases in the breath if not in the blood.

Hill: Unfortunately oxygen and nitrogen do not have i.r. absorption bands; these analyses would be much easier if they did.

REFERENCES

GROLLMAN, A. (1929). *Am. J. Physiol.*, **88**, 432.

DETERMINATION OF VOLATILE ORGANIC ANAESTHETICS IN BLOOD, GASES, TISSUES AND LIPIDS: PARTITION COEFFICIENTS

H. J. LOWE and K. HAGLER

Section of Anesthesiology,
University of Chicago Medical School,
Chicago, Illinois

THE rate of whole body anaesthetic uptake ($\dot{V}_{an}$) at constant arterial concentration can be expressed as the sum of the rates of uptake of the separate body compartments and is equal to:

$$\dot{V}_{an} = \sum \dot{Q}_i C_a e^{-Q_i t / V_i \lambda_i} \qquad (1)$$

where $\dot{Q}_i$, V_i and λ_i are the blood flow (l/m), volume (l), and tissue–blood partition coefficient of a given tissue (i). The duration of anaesthesia in minutes is designated by t. C_a is the arterial anaesthetic concentration.

One of the determinant factors in anaesthetic uptake is the solubility of the anaesthetic in the various tissue compartments, λ_i. Customarily, solubility is expressed as the ratio of the concentration of anaesthetic between two phases in equilibrium (water–gas, blood–gas or tissue–blood). This ratio, determined at a specified temperature, is referred to as the partition coefficient or Ostwald solubility coefficient (λ). A summary of existing values for various tissues and the conditions under which they were determined has been reported by Larson, Eger and Severinghaus (1962). Anaesthetics are capable of combining with proteins, haemoglobin, lipids and water. A systematic investigation of the effects of temperature, anaesthetic concentration and tissue composition on anaesthetic solubility has never been reported.

Recent developments in chromatography and detectors have greatly simplified the analysis of anaesthetics. Clinically, the volatile organic anaesthetics are employed singly or with nitrous oxide. The concentrations employed are usually in the range of 20–200 000 ppm in the inhaled vapour. Concentrations in arterial blood during anaesthesia may be several times greater than this. The high volatility and lipid solubility of these agents have resulted in numerous methods of analysis utilizing extractions, tonometry and distillations, together with thermal, flame, electron capture and ionization detectors (Adlard and Hill, 1960; Lowe, 1964*a*, *b*, *c*, 1968; Wolfson, Ciccarelli and Siker, 1966; Butler, Kelly and Zapp, 1967; Yamamura, Wakasugi and Okuma, 1960; Yokata *et al.*, 1967; Lowe and Beckham, 1964).

METHODS

The direct sample injection method employed in this laboratory has been described previously (Lowe, 1964*b*, 1968). A flame ionization detector (F and M Scientific Co., model 1609) was used. The carrier gas, nitrogen, was pre-humidified by passing through a 1·2-metre column of wet Chromosorb. The injection-port temperature was maintained between 40° and 70° C except for tissue analysis when the temperature was maintained between 100° and 110° C. The outlet of the injection-port was connected to the detector block (at 150–200° C) by means of a 0·3–0·6-metre column of Chromosorb which was moistened by injecting 1–2 ml. of water on to the column each day. One-microlitre samples of liquids or homogenates were deposited as a ring round the injection-port wall by means of a rotary motion of the syringe following penetration of the septum (Lowe, 1968). Response areas, as determined with a Disc integrator, were compared with the response areas of water or gas standards analysed under similar conditions.

Repeated gas analyses were automatically obtained every few seconds by inserting a Loenco ten-microlitre, pneumatic, solenoid-operated valve between the column and the detectors. Samples were drawn through the sample side of the valve with a to-fro syringe pump.

Tonometry

One or two drops of water, blood, lipids or tissue homogenate were equilibrated with known concentrations of various anaesthetics in oxygen at several temperatures in a Radiometer (London Co., Cleveland, Ohio, model AMT-I) tonometer. Because of technical difficulties in obtaining accurate, reproducible gas analyses, the water-gas partition coefficients, W/G, of several anaesthetics were carefully measured. Subsequent tissue–gas partition coefficients, W/G, were calculated from the ratios of the tissue response area to the water response area, T/G, without measurement of the gas concentration (that is, $T/G = T/W \times W/G$).

Tissue homogenates were prepared in glass homogenizers using one part by weight of tissue to three parts of water. When H/G, the homogenate partition coefficient, had been determined ($H/G = H/W \times W/G$), the tissue partition coefficient T/G was calculated from the formula, $T/G = (4\ H/G - 3\ W/G)$ or $W/G(4\ H/W - 3)$.

Tissue analysis

Tissue specimens obtained by biopsy or at autopsy were placed in glass or metal syringes or a Harvard tissue press with a minimum of air trapping (Fig. 1). Fifteen to 25 mg. of tissue were extruded into weighed half sections of haematocrit (approximately 60 mg.) tubes and immediately reweighed on a torsion or Cahn electrobalance. The capillary tube with the weighed tissue was inserted into an F and M solid-sample injector and introduced into the injection-port. The injector was rotated several times to grind the glass and spread the tissue over the injection-port walls, at a temperature of 100–110° C. The response area per milligramme of tissue was standardized with water standards of the anaesthetic injected in the same manner.

Tissue samples, including blood, may be extruded into flux-free solder tubing by means of appropriate adapters and the ends crushed with pliers (Lowe, 1964*c*). Specimens preserved in this way may be stored for as long as six months without loss of anaesthetic (methoxyflurane in liver and kidney excepted). The tubing holds about 0·05 ml. specimen/25 mm., and 75–100 mm. of filled tubing permits duplicate or triplicate analyses. Tissue specimens were retrieved from the tubing by cutting open one end with a sharp wire-stripper. The open end was attached to a weighed glass capillary tube with a 3 mm.–3 mm. modified Swagelok union, and the tissue expressed by crushing the distal end of the solder tube. Specimens preserved in this manner may be conveniently mailed to laboratories specializing in anaesthetic analyses (see Fig. 1).

Chromatographic lipid columns

The solid (and liquid) lipids were dissolved in ether or chloroform (0·1 g. solid or 0·1 ml. liquid/50 ml.) and added to 20 g. of 60–80-mesh glass beads. The solvent was evaporated slowly with continuous stirring on a warm water-bath. The solvent-free, lipid-coated glass beads were packed into a 60 mm. × 6·25 mm. (outside diameter) copper tube with the aid of a vibrator. Tamping of the beads was avoided to prevent excessive resistance to subsequent carrier-gas flow. Glass-wool plugs were inserted at each end of the column. The column was inserted through a water jacket consisting of plastic tubing and Swagelok T-connectors and installed between the injection-port and detector of the flame ionization detector. The water-jacket was connected to a circulating, constant-temperature water-bath.

The carrier-gas (nitrogen) flow was adjusted to give practical retention times of less than 15 min. for methoxyflurane, and the various column parameters needed to calculate the lipid-anaesthetic partition coefficient were determined.

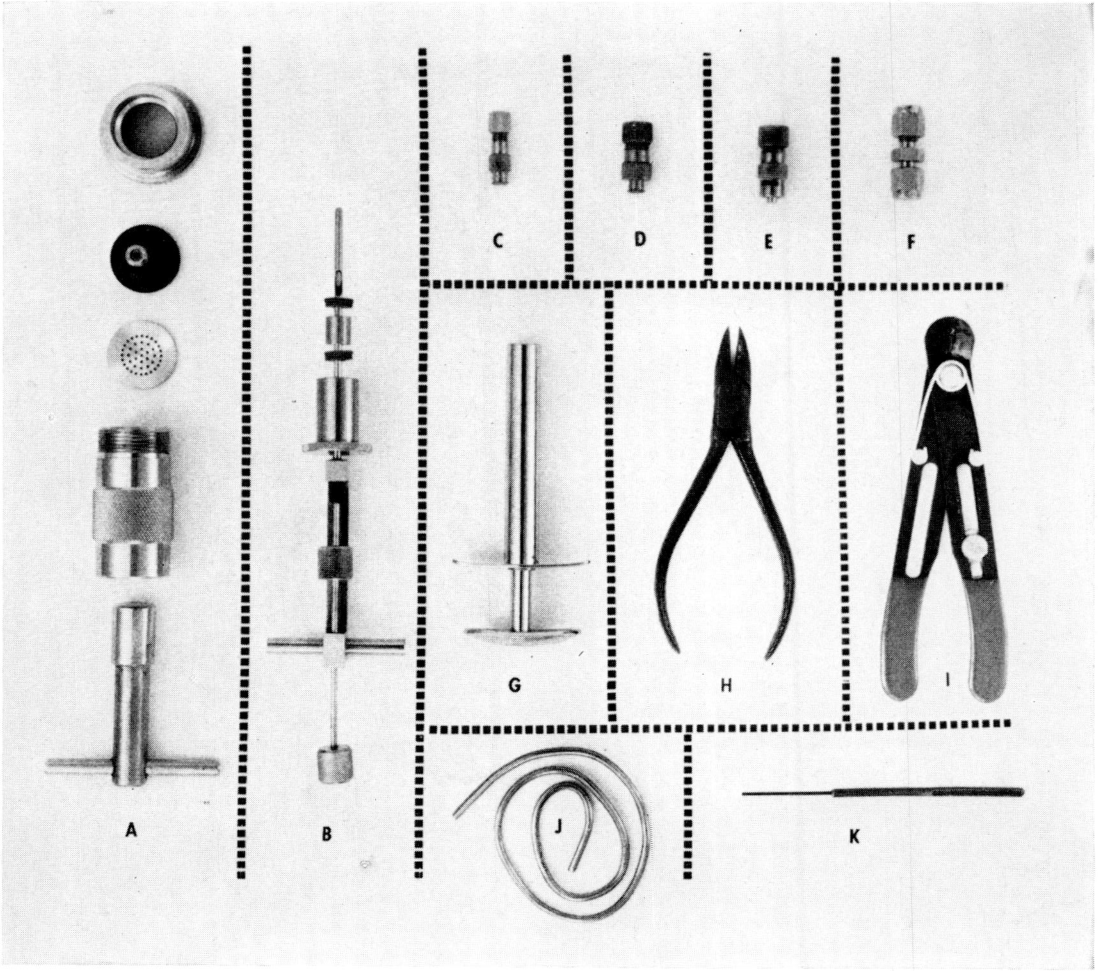

FIG. 1. Accessories for handling tissue specimens. (A) Harvard tissue press; (B) F and M solid sample injector; (C) B–D Leur-Lok adapter 610–A; (D) B–D adapter 613–A; (E) B–D adapter 614–A; (F) 0·32 mm. Swagelok union; (G) B–D metal syringe; (H) pliers; (I) wire strippers; (J) flux-free solder tubing; (K) 0·32 mm. (outer diameter) copper tube for halving haematocrit tubes.

[*to face page 88*

THEORY

Gas chromatography provides a rather simple and direct method for the determination of the partition coefficients of gas in a variety of substances (Littlewood, 1962; Szymanski, 1964; Korger, 1967). In chromatography, the partition coefficient, K, of a gas (solute) on a column packing (solvent) is related to the gas retention volume, V_R^0, by the equation:

$$K = \frac{V_R^0 - V_G^0}{V_L} = \frac{(V_R^0 - V_G^0)\,\rho_L}{w_L}$$

where: V_L = the volume occupied by the solvent (lipid) on the column; ρ_L = the density of the lipid at column temperature; w_L = the weight of the absorbent (lipid); and V_G^0 = the retention volume of an inert or non-absorbed gas.

The important parameters which must be controlled or measured in order to permit calculation of the partition coefficient are defined as follows: (1) The uncorrected retention time, t_R, is the time from sample injection to the appearance of the peak maximum of the anaesthetic compound; (2) the uncorrected retention volume, V_R, is the volume of carrier gas, measured at the column outlet pressure, p_o, required to sweep the anaesthetic from the injection-port to the detector. It is related to the carrier-gas flow rate, F_c, by the equation: $V_R = t_R . F_c$; (3) the corrected retention volume, V_R^0, is the value V_R corrected for the gas compressibility resulting from the pressure gradient across the column, that is,

$$V_R^0 = V_R \frac{3}{2}\left[\frac{(p_i/p_o)^2 - 1}{(p_i/p_o)^3 - 1}\right],$$

where p_i and p_o are the column inlet and outlet pressures (absolute); and (4) the total gas volume, V_G^0, is the corrected retention volume of a non-absorbed gas (that is, oxygen).

When the parameters F_c, p_o, p_i, V_G^0 and ρ_L for a particular lipid column at a given temperature are known, the partition coefficient of an anaesthetic in the lipid may be determined from the time needed for the anaesthetic to pass through the chromatograph. For accurate results the anaesthetic vapour and the lipid phase should be in equilibrium throughout the column, and the anaesthetic peak should appear as a symmetrical Gaussian curve. This is best accomplished by minimal lipid packing consistent with avoiding exposure of absorbing sites on the lipid support (glass beads), and by injecting small quantities of anaesthetic vapour in order to achieve thermodynamic equilibrium at near infinite dilution.

Under these conditions, and over a limited temperature range, the enthalpy of vaporization, ΔH, of the anaesthetic vapour from the lipid phase may be related to the lipid–gas partition coefficient by the Gibbs–Helmholtz equation (Glasstone, 1940):

$$\log \frac{K_2}{K_1} = \frac{\Delta H}{2 \cdot 3R} \left[\frac{1}{T_2} - \frac{1}{T_1}\right] \tag{2}$$

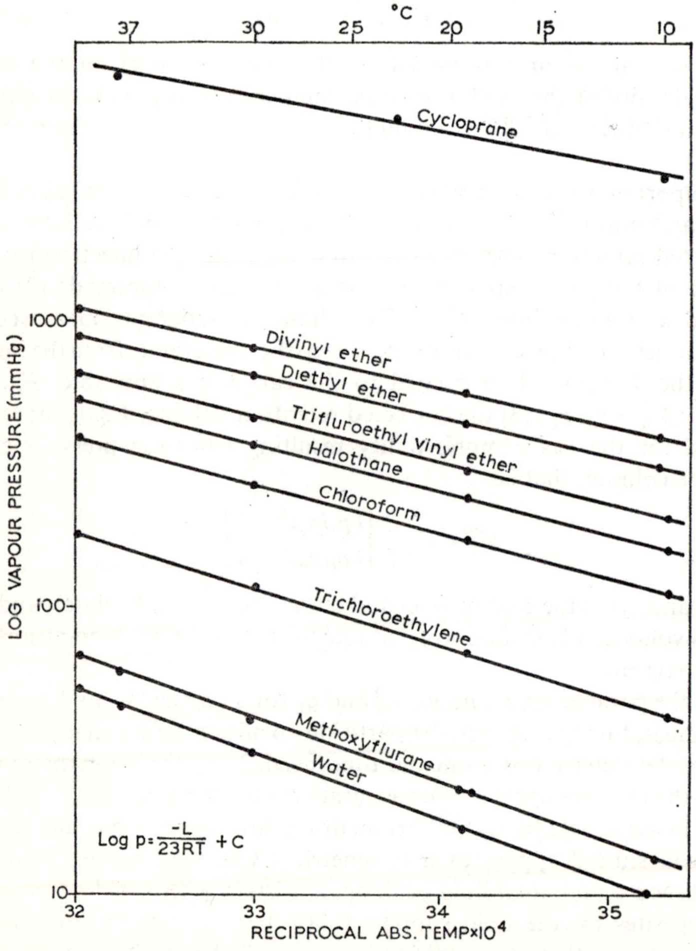

FIG. 2. Graph of log of vapour pressure plotted against reciprocal absolute temperature. Top abscissa indicates °C.

where R is the gas constant, 8·2 J/mole/°C, and K_2 and K_1 are the respective lipid partition coefficients at absolute temperatures T_2 and T_1.

In addition, it is known that the vapour pressure, p, of an anaesthetic varies with the absolute temperature, T, according to the Clausius–Clapeyron equation (Glasstone, 1940):

$$\log \frac{p_2}{p_1} = -\frac{L}{2{\cdot}3R}\left[\frac{1}{T_2} - \frac{1}{T_1}\right] \qquad (3)$$

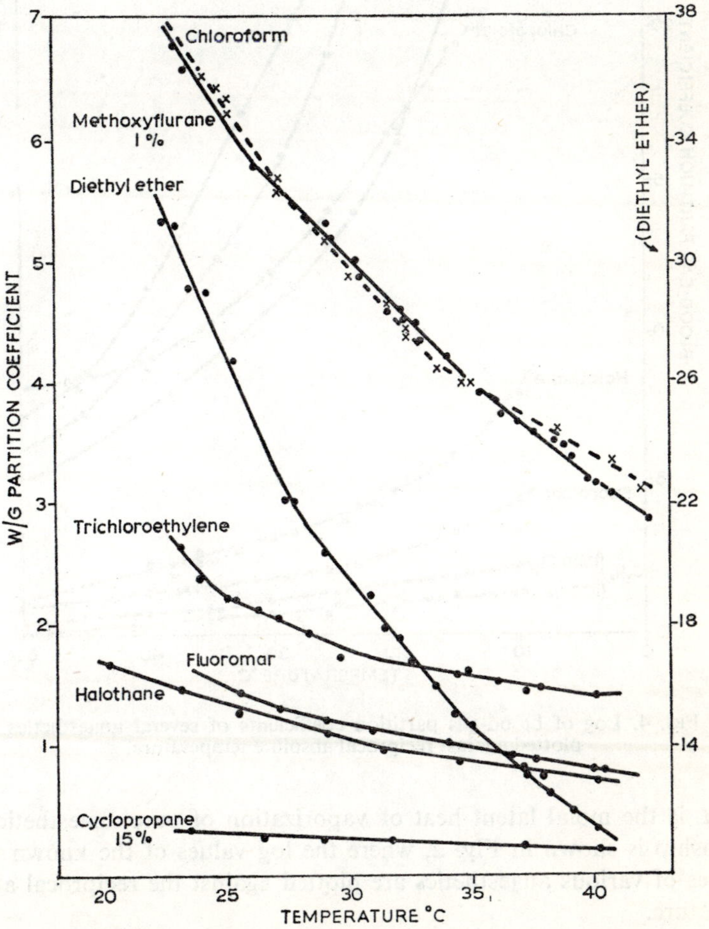

FIG. 3. Water–gas partition coefficients. W/G partition coefficients of ether are listed on right ordinate.

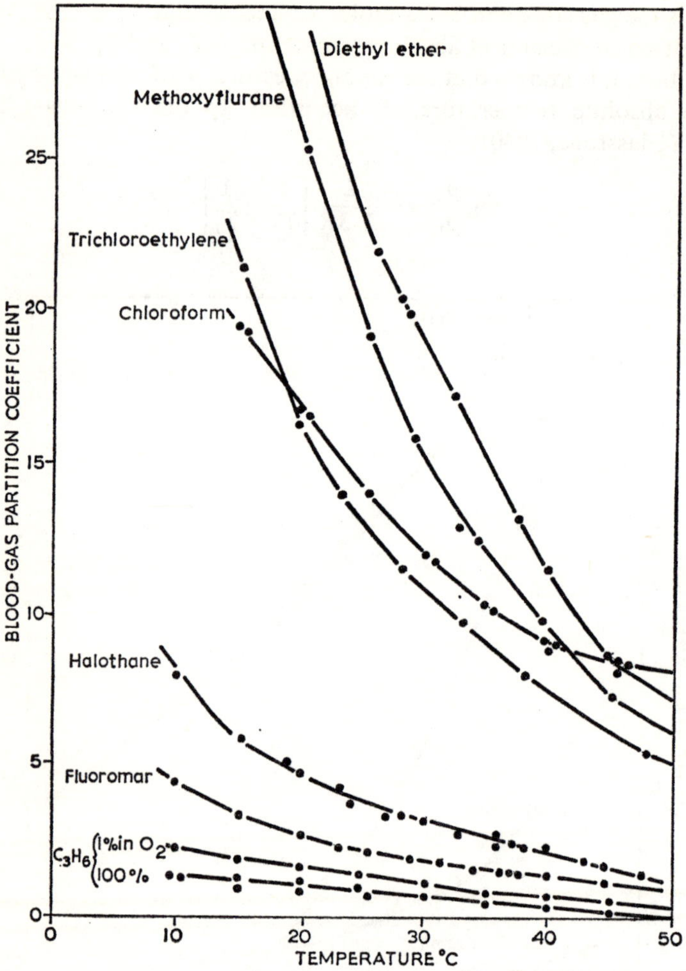

FIG. 4. Log of blood-gas partition coefficients of several anaesthetics plotted against reciprocal absolute temperature.

where L is the molal latent heat of vaporization of pure anaesthetic. This relationship is shown in Fig. 2, where the log values of the known vapour pressures of various anaesthetics are plotted against the reciprocal absolute temperature.

Over a given temperature range where ΔH and L are relatively constant, it follows that the solubility of an anaesthetic in a lipid is inversely proportional to the vapour pressure, p, of the anaesthetic at a given temperature—equation (3) divided by equation (4):

$$\log \frac{K_2}{K_1} = -\frac{\Delta H}{L} \log \frac{p_2}{p_1} \tag{4}$$

If, as has been assumed by E. I. Eger and his group (Saidman *et al.*, 1967), the concentration of anaesthetic dissolved in the "material" constituting the

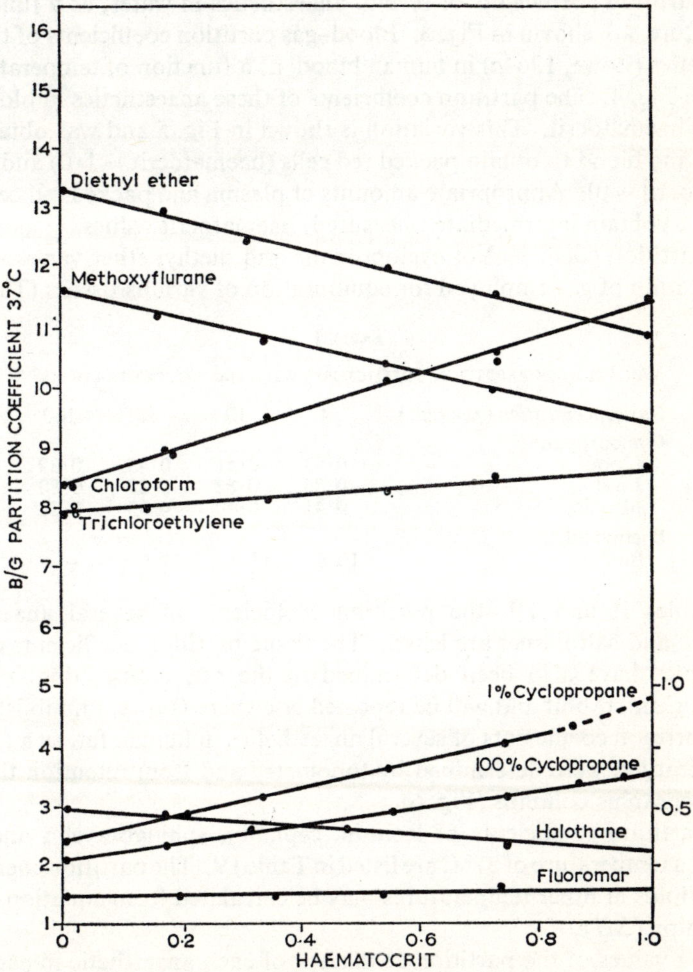

FIG. 5. Effect of haematocrit on blood-gas partition coefficients of anaesthetics at 37°C.

site of anaesthetic action is the same at different temperatures when equipotent anaesthetic gas concentrations, *MAC*, are administered, it follows that

$$\log\frac{MAC_1}{MAC_2} = \log\frac{K_2}{K_1} = \frac{\Delta H}{L}\log\frac{p_1}{p_2} = \frac{\Delta H}{2\cdot 3R}\left[\frac{1}{T_2} - \frac{1}{T_1}\right] \qquad (5)$$

RESULTS

The partition coefficients of several anaesthetics in water, as a function of temperature, are shown in Fig. 3. Blood–gas partition coefficients of the same anaesthetics (Lowe, 1964*a*) in human blood, as a function of temperature, are shown in Fig. 4. The partition coefficients of these anaesthetics in blood vary with the haematocrit. This variation is shown in Fig. 5 and was obtained by centrifuging blood to obtain packed red cells (haematocrit = 1·0) and plasma (haematocrit = 0). Appropriate amounts of plasma and packed red cells were remixed to obtain intermediate (measured) haematocrit values.

The partition coefficient of cyclopropane and diethyl ether varies with the concentration of gas employed for equilibration of various tissues (Table I).

TABLE I

VARIATION IN PARTITION COEFFICIENTS WITH GAS CONCENTRATIONS

Gas concentration (per cent)	4	10	50	100
Cyclopropane:				
Blood	0·53	0·51	0·45	0·39
Liver	0·88	0·88	0·88	0·80
Muscle	0·41	0·40	0·33	0·22
Diethyl ether:				
Blood	10·8	11·8	15·0	—

In Tables II and III, the partition coefficients of several anaesthetics in human and calf tissues are listed. The tissue partition coefficients of these anaesthetics have also been determined in the rat, mouse, dog, monkey, guinea-pig and rabbit and will be reported elsewhere (Lowe, unpublished).

The partition coefficients of several anaesthetics in human fat, as a function of temperature, were determined by tonometry and from retention times on chromatographic columns (Fig. 6).

The partition coefficients of lecithin, cephalin, sphingomyelin and cholesterol at a temperature of 37° C are listed in Table IV. The partition coefficients of these lipids at other temperatures may be calculated from equation (2) and the enthalpy (ΔH).

The log values of the partition coefficient of each anaesthetic in each lipid at 37° C were plotted against the logarithms of the minimum concentrations required for anaesthesia, *MAC*, and are shown in Fig. 7.

TABLE II

HUMAN TISSUE-GAS PARTITION COEFFICIENTS (37° C)
(HUMAN NO. 2)

Tissue	*Cyclopropane*	*Fluroxene*	*Diethyl ether*	*Halothane*	*Chloroform*	*Trichloro-ethylene*	*Methoxy-flurane*	*Divinyl ether*
Water	0·240	0·92	13·70	0·86	3·91	1·51	3·49	1·45
Blood	0·565	—	—	2·56	—	—	—	2·22
Liver	0·656	1·96	12·20	7·24	16·60	17·30	24·84	2·56
Kidney	0·352	1·44	12·88	4·02	10·28	10·48	12·12	1·68
Brain (white)	0·848	2·52	13·52	8·22	20·12	34·00	30·68	3·6
Brain (grey)	0·432	1·48	12·52	4·08	11·60	16·92	16·32	4·0
Muscle	0·656	1·64	11·80	6·72	15·84	18·96	24·68	2·48
Pancreas	0·968	3·24	14·70	12·24	26·16	47·48	50·52	5·8
Thyroid	0·656	1·52	11·48	4·96	11·08	17·28	21·32	2·11
Adrenal	2·58	6·20	22·24	33·62	46·96	129·88	172·0	12·0
Fat	13·0	53·8	73·5	182·0	394·0	634·0	820·0	59·3
Spleen	0·392	1·04	9·28	2·92	8·76	7·40	10·04	1·68
Bile	0·752	2·36	13·35	2·74	8·72	10·04	10·25	4·16

TABLE III

TISSUE–GAS PARTITION COEFFICIENTS AT 37° C (CALF)

Tissue	Cyclopropane		Fluroxene		Diethyl ether		Halothane		Chloroform		Trichloro-ethylene		Methoxy-flurane		Divinyl ether	
	*Calculated**	*Observed*	*Calculated*	*Observed*	*Calculated*	*Observed*	*Calculated*	*Observed*	*Calculated*	*Observed*	*Calculated*	*Observed*	*Calculated*	*Observed*	*Calculated*	*Observed*
H_2O	—	0·212	—	0·96	—	13·7	—	0·88	—	3·85	—	1·51	—	3·50	—	1·62
Blood	—	0·480	—	1·19	—	11·9	—	2·87	—	8·57	—	6·70	—	8·40	—	2·46
Brain	0·68	0·670	1·73	1·70	11·6	10·5	5·5	4·86	18·0	20·90	23·0	18·90	24·0	23·80	3·26	3·09
Muscle (0·65% fat)	0·48	0·450	1·50	1·34	10·4	10·2	2·3	2·20	7·6	7·05	6·3	6·27	10·0	9·20	1·90	2·14
Liver (2·0% fat)	0·64	0·738	2·0	2·07	12·8	13·4	6·8	6·40	15·9	16·20	25·0	29·30	26·3	25·90	2·70	3·70
Thymus	—	0·390	—	0·80	—	10·5	—	1·44	—	3·45	—	8·47	—	6·80	—	2·03
Thyroid	—	0·414	—	0·80	—	9·4	—	1·20	—	4·45	—	5·47	—	6·90	—	1·83
Kidney	0·42	0·626	1·90	1·58	10·8	10·9	4·1	4·35	7·6	7·50	13·4	14·90	17·3	18·90	2·80	3·39
Spleen	—	0·474	—	0·86	—	10·9	—	3·00	—	6·95	—	8·17	—	10·70	—	2·46
Adrenal	—	0·660	—	1·36	—	10·8	—	1·62?	—	14·30	—	16·70	—	17·60	—	1·86
Fat (extracted)	—	12·3	—	48·8	—	66·0	—	181·0	—	515·0	—	564·0	—	690·0	—	64·7
Lung	0·42	—	1·50	—	11·5	—	4·1	—	11·0	—	14·0	—	18·6	—	2·45	—

* See Kaucher and co-workers (1943).

TABLE IV

ENTHALPY, HEAT OF VAPORIZATION AND PARTITION COEFFICIENTS OF ANAESTHETICS IN ANIMAL LIPID

	Lipid										
	Human fat		α-Lecithin		Cephalin		Sphingomyelin		Cholesterol		Latent heat of vaporization
Anaesthetic	K	ΔH	K	ΔH	K	ΔH	K	ΔH	K	ΔH	
Cyclopropane	12	1·00	8·2	3·22	6·5	2·74	1·0	0·92	1·2	2·76	−3·4
Trifluroethyl, vinyl ether	36	4·96	20	7·45	15	6·14	—	—	2·7	4·70	−6·95
Divinyl ether	46	4·70	28	5·30	22	5·94	7·3	2·48	7·0	7·60	−5·30
Diethyl ether	57	5·95	30	6·25	25	5·80	11	2·34	20	11·50	−5·65
Halothane	186	8·05	94	7·65	64	8·25	20	2·85	8·0	2·76	−7·13
Chloroform	400	8·16	190	7·24	140	7·77	49	4·60	38	9·45	−7·45
Trichloroethylene	740	8·25	300	7·30	260	8·45	100	6·67	76	10·85	−8·60
Methoxyflurane	900	8·75	365	8·16	260	9·45	82	5·75	42	13·10	−9·55

K = partition coefficient (37° C); ΔH = kcal./mole, where 1 kcal. = 4·19 kJ.

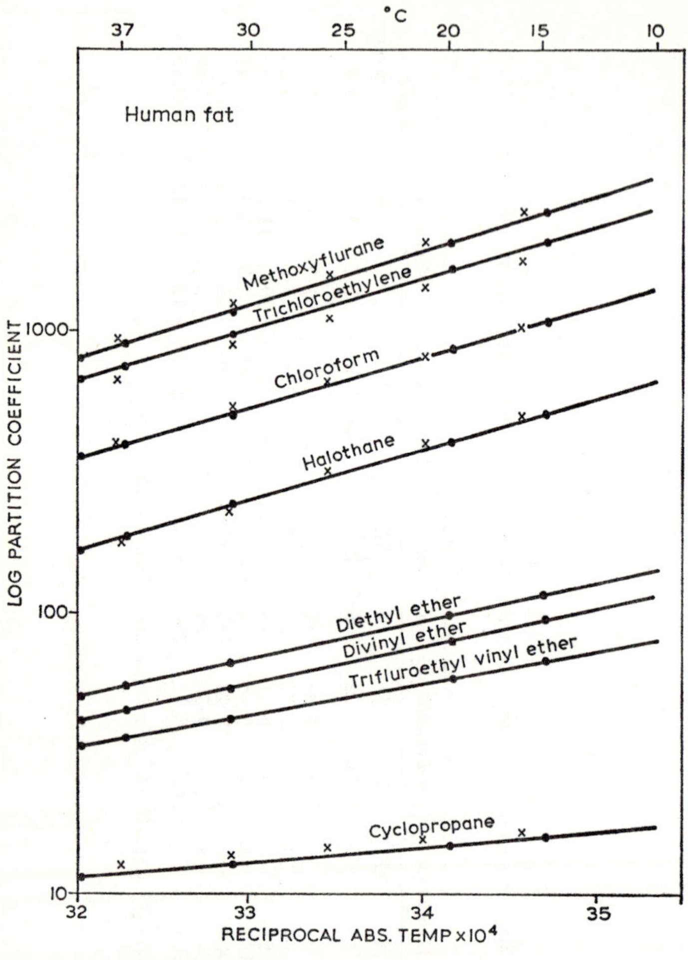

FIG. 6. Log of fat-gas partition coefficients in man plotted against reciprocal absolute temperature. Values obtained by tonometry: ×; by retention times on chromatographic columns: ●.

The partition coefficients of white and grey matter of the brain were calculated from the published composition (West and Todd, 1951; Long, 1961) of these tissues and the observed lipid partition coefficients (Table IV). These calculated partition coefficients are compared with *in vitro* measurements of whole brain and grey and white matter homogenates in Table V.

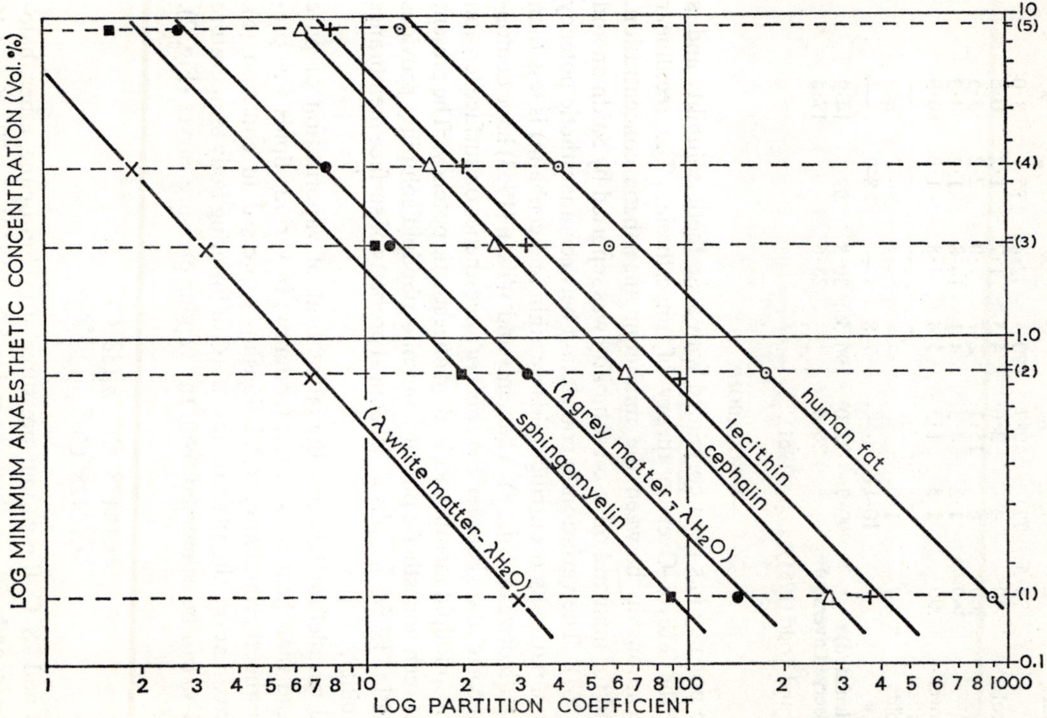

FIG. 7. Rectilinear relationship between minimum anaesthetic concentration (*MAC*) and solubility of anaesthetics in phospholipids and white and grey matter of brain. Partition coefficients of grey matter have been multiplied by 10. Product of *MAC*·λ (grey matter) for each anaesthetic varies between 2·32 and 2·60, or ±5% of mean. Horizontal broken lines represent *MAC* as follows: (A) methoxyflurane: 0·16; (B) halothane: 0·77; (C) ether: 1·94; (D) fluroxene: 3·4; and (E) cyclopropane: 9·2. *MAC* measured as ml. anaesthetic/100 ml. alveolar gas.

TABLE V

SOLUBILITY OF METHOXYFLURANE IN BRAIN (37°*C*)

Substance	λ	*Whole brain* %	*Whole brain* λ	*White matter* %	*White matter* λ	*Grey matter* %	*Grey matter* λ
Water	3·5	77	2·7	70	2·45	85	3·0
Cholesterol	40	4	2·6	4·3	1·7	1·2	0·5
Cephalin	260	6	15·1	2·7	7·0	2·2	5·5
Lecithin	360	1·2	4·3	4·1	14·8	1·2	4·3
Sphingomyelin	90	1·5	1·3	5·4	4·8	1·0	0·9
Cerebrosides	?	—	—	—	—	—	—
Sulpholipids	?	1·0	—	—	—	—	—
Proteins	—	10–11	—	7–8	—	8–9	—
Total (calculated)*		99·7	27·0	96·5	30·8	99·6	14·2
Total (observed average)			26·4		33·6		17·5

* West and Todd (1951); Long (1961).

DISCUSSION

Anaesthetic potency has been correlated with anaesthetic solubility and is the basis of the Meyer–Overton theory of anaesthesia. The rectilinear logarithmic relationship between the minimum anaesthetic concentration, *MAC*, required for man and olive-oil solubility was reported by Saidman and co-workers (1967). The variation in correlation between anaesthetic potency and solubility in naturally occurring lipids (lecithin and cephalin) is less than that observed with olive oil (Fig. 7). Clements and Wilson (1963) have demonstrated a direct correlation between the effect of anaesthetics on surface tension of lipid and phospholipid membranes and on anaesthetic potency. The present correlation between anaesthetic potency and phospholipid solubility provides a direct correlation between a physicochemical property of surface membranes and lipid solubility.

The general parallelism between the latent heat of vaporization of pure anaesthetics and their heat of solution (enthalpy) in various lipids was predictable from the thermodynamic relationship expressed in equation (5). As a first approximation, the effect on the *MAC* of lowering body temperature from 38 to 28°C can be calculated from the vapour pressure curves (Fig. 2), that is:

$$\frac{MAC(38°\text{C})}{MAC(28°\text{C})} = \frac{p_1(38°\text{C})}{p_2(28°\text{C})}$$

Between 38 and 28°C, the vapour pressures of cyclopropane, ether, halothane, chloroform and methoxyflurane are decreased by 21, 28, 32, 36 and 46 per cent respectively. These calculated decreases agree with those reported by Eger, Saidman and Brandstater (1965) for dogs, and by Cherkin and Catchpool (1964) for goldfish.

The partition coefficient of halothane in brain homogenate, as a function of temperature, was reported by Han and Helrich (1966). These data have a solubility–temperature slope ($\Delta H = 7{\cdot}95$ kcal./mole) intermediate to that for halothane in lecithin or cephalin (7·65–8·25 kcal./mole) (Table IV).* The enthalpy of solution for halothane in water (rectilinear logarithmic plot of Fig. 3) was 9·0 kcal./mole and agrees with that reported by Eger, Saidman and Brandstater (1965); however, our value for halothane in olive oil (7·0 kcal./mole) is less than the 9·0 kcal./mole reported by these authors. The enthalpy of absorption of cyclopropane in water of 5·7 kcal./mole agrees with the enthalpy of absorption of cyclopropane by dogs (Eger, Saidman and Brandstater, 1965).*

A variety of physical techniques have demonstrated that phospholipids are capable of existing as liquid crystals at body temperature (Chapman, 1967). Regardless of the physical state of the phospholipids on the glass-bead support, the derived partition coefficients were independent of the amounts of vapour injected. The validity of application of these values to biological systems is demonstrated by their ability to account for the solubility of the various anaesthetics in brain (Table V). The solubilities of the remaining anaesthetics in white and grey matter were similarly calculated, using the brain composition shown in Table V and the anaesthetic partition coefficients listed in Table IV, and are shown in Table VI.

TABLE VI

CALCULATED PARTITION COEFFICIENTS OF ANAESTHETICS IN WHITE AND GREY MATTER OF BRAIN

	White matter		*Grey matter*	
Anaesthetic	*calculated*	*observed*	*calculated*	*observed*
Cyclopropane	0·78	0·848	0·44	0·432
Fluroxene	2·301	2·52	1·45	1·48
Divinyl ether	3·47	4·02	2·25	2·82
Diethyl ether	12·9	13·52	12·82	12·52
Halothane	7·51	8·22	3·70	4·08
Chloroform	18·63	20·12	9·63	11·60
Trichloroethylene	29·0	34·0 (31·2)	12·5	16·2 (13·2)

In the white matter of the brain the calculated partition coefficients average about 92 per cent (87–95 per cent) of the observed *in vitro* partition coefficients. Brain also contains cerebroside, galactosides, sulpholipid and traces of triglycerides in amounts capable of accounting for the observed differences without invoking protein binding, although many of these lipids are associated with proteins. When the observed partition coefficients of the white and grey

* One kilocalorie = 4·19 kilojoule.

matter of the brain for each anaesthetic are corrected for the contribution of water to the total partition coefficient (that is, λ (grey matter) $- \lambda_{H_2O} \cdot \%H_2O$), and the log of the remaining partition coefficient (brain lipid) is plotted against the log of the *MAC*, a rectilinear relationship similar to that of pure lipid is obtained (Fig. 7).

In general, the solubility or partition coefficient of an anaesthetic in a given tissue may be calculated with reasonable accuracy from the lipid and water partition coefficients if the composition of the tissue is known.

$$\lambda_t = (\lambda . f)_{H_2O} + (\lambda . f)_{cephalin} + (\lambda . f)_{lecithin} + (\lambda . f)_{sphingomyelin} + (\lambda . f)_{cholesterol} + (\lambda . f)_{triglycerides}$$

where $(\lambda . f)$ is the product of the partition coefficient and the fractional composition of the particular component. These calculations are shown in Table III. The agreement with observed values is very good for organs of constant composition (brain). For other organs (liver and muscle), the simultaneous equations were solved to determine the percentage of fat in the specimen. The tissue–gas partition coefficients for lung were calculated from the reported composition (Kaucher *et al.*, 1943). The calculated values for cyclopropane are low due to the interaction of this gas with proteins (Table I) (Featherstone *et al.*, 1961). Although interaction of other anaesthetics cannot be completely ruled out, such interaction does not appear to contribute significantly to the tissue solubility of anaesthetics.

SUMMARY

The water–gas and blood–gas partition coefficients of cyclopropane, trifluoroethyl, vinyl ether, divinyl ether, diethyl ether, halothane, chloroform, trichloroethylene and methoxyflurane were determined between temperatures of 10 and 40° C and as a function of the blood haematocrit. The blood–gas and tissue–gas partition coefficients of cyclopropane increase with decreasing gas concentration. Tissue–gas partition coefficients of human and calf tissues were determined at 37° C. The anaesthetic partition coefficients in human fat, cephalin, lecithin, sphingomyelin and cholesterol were determined between 10 and 40° C from their retention time on chromatographic columns and, where possible, by means of tonometry. The enthalpy of solution of anaesthetic in each lipid was obtained from the Gibbs–Helmholtz equation and was constant between temperatures of 10 and 40° C. The observed partition coefficients of these lipids were utilized to calculate the anaesthetic partition coefficients in human brain and calf tissues from the known lipid composition of these tissues. These values compared favourably with those obtained by tonometry.

The inverse linear logarithmic relationship between anaesthetic potency and anaesthetic solubility in phospholipid and brain tissue supports the lipoid theory of anaesthesia.

Acknowledgement

Supported in part by US Army Medical Research and Development Command Contract DA-DA-17-67-C-7052.

REFERENCES

ADLARD, E. R., and HILL, D. W. (1960). *Nature, Lond.*, **186**, 1045.

BUTLER, R. A., KELLY, A. B., and ZAPP, J. (1967). *Anesthesiology*, **28**, 760–763.

CHAPMAN, D. (1967). In *Ordered Fluids and Liquid Crystals*, pp. 157–166, ed. Porter, R. S. Washington, D.C.: American Chemical Society.

CHERKIN, A., and CATCHPOOL, J. F. (1964). *Science*, **144**, 1460–1462.

CLEMENTS, J. H., and WILSON, K. M. (1963). *International Anesthesiology Clinics*, Vol. 1, pp. 969–979. Boston: Little, Brown.

EGER, E. I., SAIDMAN, L. J., and BRANDSTATER, B. (1965). *Anesthesiology*, **26**, 764–770.

FEATHERSTONE, R., MUELBAECHER, C., DEBON, F., and FORSAITH, J. (1961). *Anesthesiology*, **22**, 977–981.

GLASSTONE, S, (1940). *Textbook of Physical Chemistry*. Princeton: Van Nostrand.

HAN, Y. H., and HELRICH, M. (1966). *Anesth. Analg. curr. Res.*, **45**, 775–780.

KAUCHER, M., GALBRAITH, H., BUTTON, Y., and WILLIAMS, H. H. (1943). *Archs Biochem.*, **3**, 203.

KORGER, B. L. (1967). *Analyt. Chem.*, **39**, 24*A*.

LARSON, C. P., EGER, E. I., and SEVERINGHAUS, J. H. (1962). *Anesthesiology*, **23**, 686–689.

LITTLEWOOD, A. B. (1962). *Gas Chromatography: Principles, Techniques and Applications*. New York: Academic Press.

LONG, C. (1961). *Biochemist's Handbook*, pp. 641–646. Princeton: Van Nostrand.

LOWE, H. J. (1964*a*). *J. Gas Chromat.*, **2**, 380–384.

LOWE, H. J. (1964*b*). *Anesthesiology*, **25**, 808–814.

LOWE, H. J. (1964*c*). *Anesthesiology*, **25**, 566.

LOWE, H. J. (1968). In *Gas Chromatography*, pp. 194–209, ed. Bender, S. R., and Kroman, H. S. New York: Grune and Stratton.

LOWE, H. J., and BECKHAM, L. M. (1964). In *Biomedical Applications of Gas Chromatography*, pp. 307–323, ed. Szymanski, H. A. New York: Plenum Press.

SAIDMAN, L. J., EGER, E. I., MUNSON, E. S., BABAD, A. A., and MUALLEM, M. (1967). *Anesthesiology*, **28**, 994–1002.

SZYMANSKI, H. A. (1964). In *Biomedical Applications of Gas Chromatography*, pp. 1–38, ed. Szymanski, H. A. New York: Plenum Press.

WEST, E. S., and TODD, W. R. (1951). In *Textbook of Biochemistry*, p. 422. New York: Macmillan.

WOLFSON, B., CICCARELLI, H. E., and SIKER, E. S. (1966). *Br. J. Anaesth.*, **38**, 29–34.

YAMAMURA, H., WAKASUGI, B., and OKUMA, Y. (1960). *Mass Spectroscopy*, **14**, 68–75.

YOKATA, T., HITOMI, Y., OHTA, K., and KOSAKA, F. (1967). *Anesthesiology*, **28**, 1064–1073.

DISCUSSION

Scott: How do you decide which programme to use for a particular patient?

Lowe: Assuming a constant arterial concentration of halothane, the rate of anaesthetic uptake of each organ compartment is calculated from anaesthetic solubility, organ weight and blood flow. The whole-body (100 kg.) uptake was obtained by summing the uptakes of individual organs. The dose-time requirement was transferred to a Data Trak potentiometric curve-follower which, in turn, operated a Harvard syringe pump at the predicted rates. A rheostat connecting the Data Trak to the Harvard pump permitted dialling a programme appropriate to the patient's weight.

Scott: What adjustments do you make for age?

Lowe: For patients who are more than 60 years old we usually reduce the appropriate predicted dose of anaesthetic by 10 per cent and, for those over 80 years, by 20 per cent.

Scott: I am enthusiastic about this system. This is the way progress in instrumentation must go.

Lowe: We have also made some interesting incidental observations. For evaluating its usefulness we try to use healthy, normal individuals, between 20 and 50 years of age, who are undergoing elective surgery. With the methoxyflurane (Penthrane) programme the predicted blood levels occur and are maintained according to our mathematical model. But, for halothane, blood levels are consistently lower than predicted—the rate of uptake of this anaesthetic is disproportionate to the concentration administered. This is a characteristic of halothane itself, which is a potent vasodilator whose administration causes readjustments in blood flow to the various organs. Part of the increased amounts of halothane that have to be given to maintain predicted blood levels are due to variable distribution within the fat compartment itself. Our model shows fat as a single compartment but, in practice, blood flow to mesenteric, omental and subcutaneous fat is entirely different. Biopsies from mesenteric and omental fat, at different intervals after a constant arterial concentration of halothane has been established and maintained, have shown that blood flow may be as high as 100 ml./kg. for mesenteric fat, but only about 18 ml./kg. for subcutaneous fat. This division of the total fat compartment into three sub-compartments has increased our understanding of the differences in uptake requirements by different individuals.

Purves: I see a gap between the theoretical determination of partition coefficients in various fat compartments and the problem set by an actual patient who may be both old and fat and show most bizarre blood flow and fat distribution. What sort of correlation have you found between predicted and observed values?

Lowe: We have now used the closed-circuit methoxyflurane programme in 30 patients. With no exceptions, by simply turning on the switch and then shutting it off at the proper time, we have obtained blood levels within 10 per cent of the predicted values. Such reliability, of course, is due to the fact that methoxyflurane does not significantly disturb blood flow to any organ.

Purves: I am astonished that your predictions are so reliable.

Payne: Another great advantage with this system is that no gas escapes to the atmosphere. With most of the anaesthetic techniques in common use a substantial amount of anaesthetic gas is lost in this way.

Lowe: I agree. Nevertheless, this sort of programming is not limited to closed systems but can also be used with semi-closed systems provided the amounts of anaesthetic vapour escaping from the circuit every minute are added to the programme.

Payne: This is only true if alveolar ventilation is maintained at a constant value. In patients breathing spontaneously the variations in alveolar ventilation will lead to irregular uptake.

Lowe: Right. A closed system such as the one I described is not subject to minor variations in ventilation or cardiac output. The increased ventilation in a semi-closed system leads to the absorption of more than the predicted amount of anaesthetic. In the closed system the amount of anaesthetic present at any one time is limited; if one accelerates the uptake momentarily by increasing ventilation, the circuit concentration and uptake drop automatically because anaesthetic is trickling in at a fixed rate.

Cervenko: How do you decide on the oxygen requirements of a patient during closed-circuit anaesthesia?

Lowe: When the system is programmed for an alveolar methoxyflurane concentration of 0·2 per cent I assume that the patient will be receiving 99·8 per cent oxygen. This is the oxygen concentration. We use a spirometer bellows to measure oxygen consumption. The height of the bellows is maintained at a constant level, which is regulated by a flowmeter micrometer needle-valve.

Robinson: Do you wash nitrogen out of the circuit first?

Lowe: Yes. One interesting finding from our studies on the programming of anaesthetics is that the predicted dose requirements, on a weight basis, for halothane and methoxyflurane are identical; so we use the same programme for both.

Geddes: Your approach to automated anaesthesia is much more promising than previous attempts based on electroencephalographic patterns. Kiersey, Falconer and Bickford (1954) and Bellville, Artusio and Bulmer (1954), who did this work, were troubled with external electrical discharges picked up from nearby lifts, and other background interference. We have approached the problem of the whole-body distribution of halothane in a rather different way from you. In collaboration with workers at the Brookhaven National Laboratories (Geddes *et al.*, 1969), we gave small quantities of [^{82}Br]halothane to humans by inhalation and plotted the whole-body distribution using a unique, 54-channel, whole-body counter.

Following neutron irradiation of commercially available halothane in an atomic reactor, chemical bonds are ruptured and the material is grossly contaminated with impurities. Fig. 1 shows these impurities, which are present

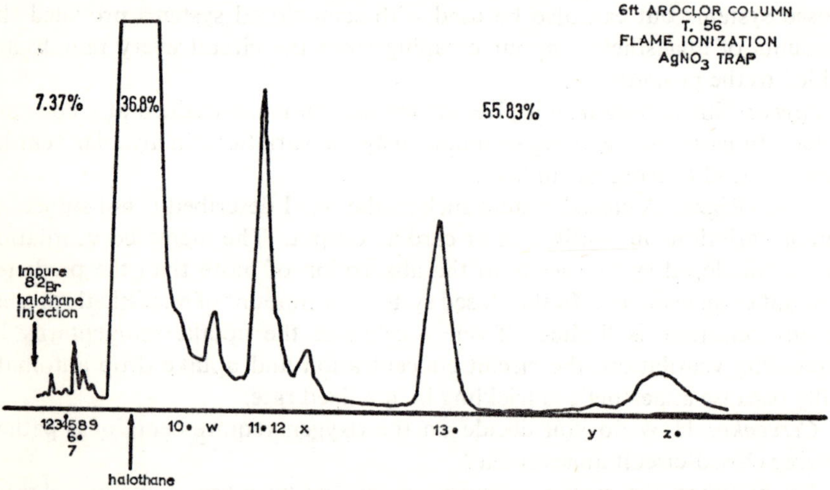

FIG. 1. Following irradiation of halothane by neutrons numerous impurities were present before and after the halothane peak. [^{82}Br] was present in peaks 4, 6, 10, 11, 13 and Z (dotted); compounds 10, 11, 13 and possibly Z are dibromo compounds which accounts for the high percentage of radioactivity recovered after the halothane peak. Percentages of impurities detected before and after halothane peak are 7·31 and 55·83 respectively.

before and after the halothane peak. Some of the impurities are dibromo compounds and must be removed before [^{82}Br]halothane can be used in man.

Two methods of checking the purity of [^{82}Br]halothane have been explored: (A) the use of a commercially available sample changer (Hamilton Co.) which is attached by a heated line to a gas chromatograph with a sample splitter. The different fractions coming off the column were collected in U-tubes containing sand, which were placed in liquid nitrogen in a Dewar flask. The tubes were then monitored for [^{82}Br]-containing compounds using a thallium-activated, sodium iodide scintillation detector (Fig. 2*a*). Fig. 2*b* shows that in a partially

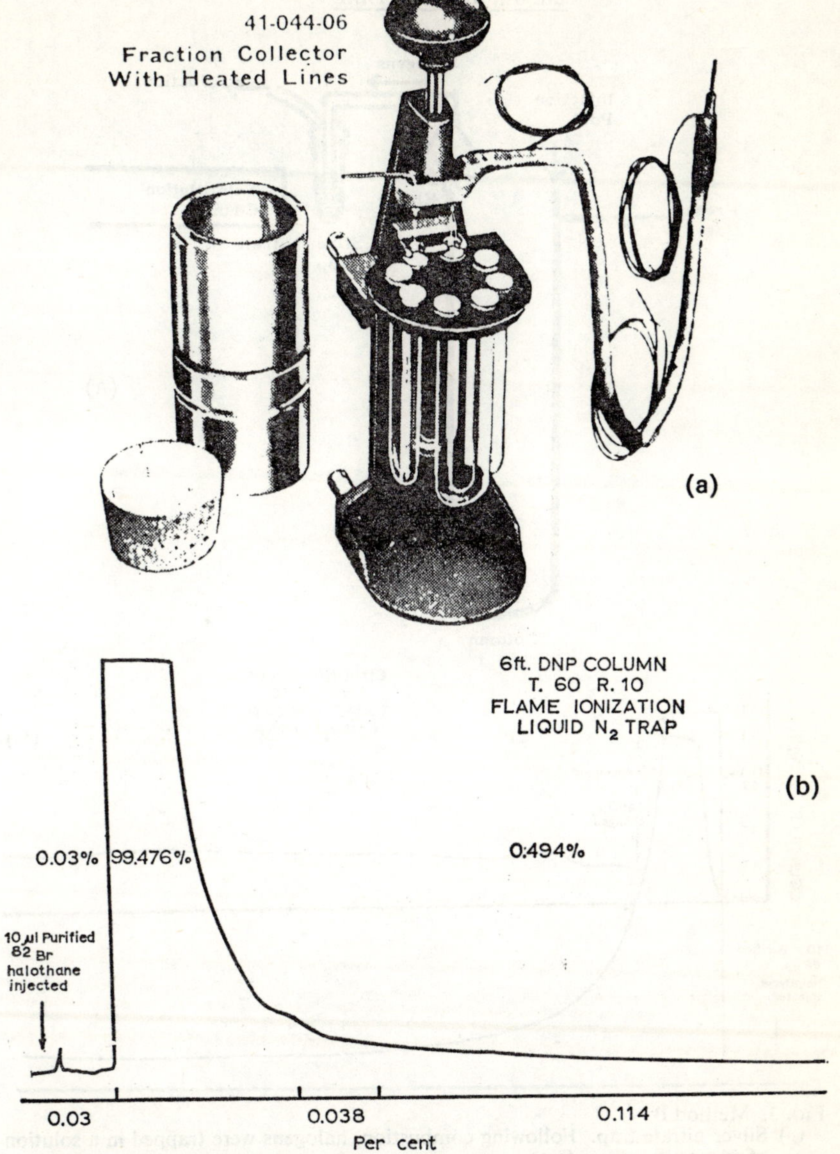

FIG. 2. Method A

(*a*) Fraction collector. Impurities were trapped on sand in U-tubes placed in liquid nitrogen, and radioactivity in the tubes was counted (see text).

(*b*) Trace shows that some impurities (0·03% before and 0·494% after halogen peak) were still present after purification.

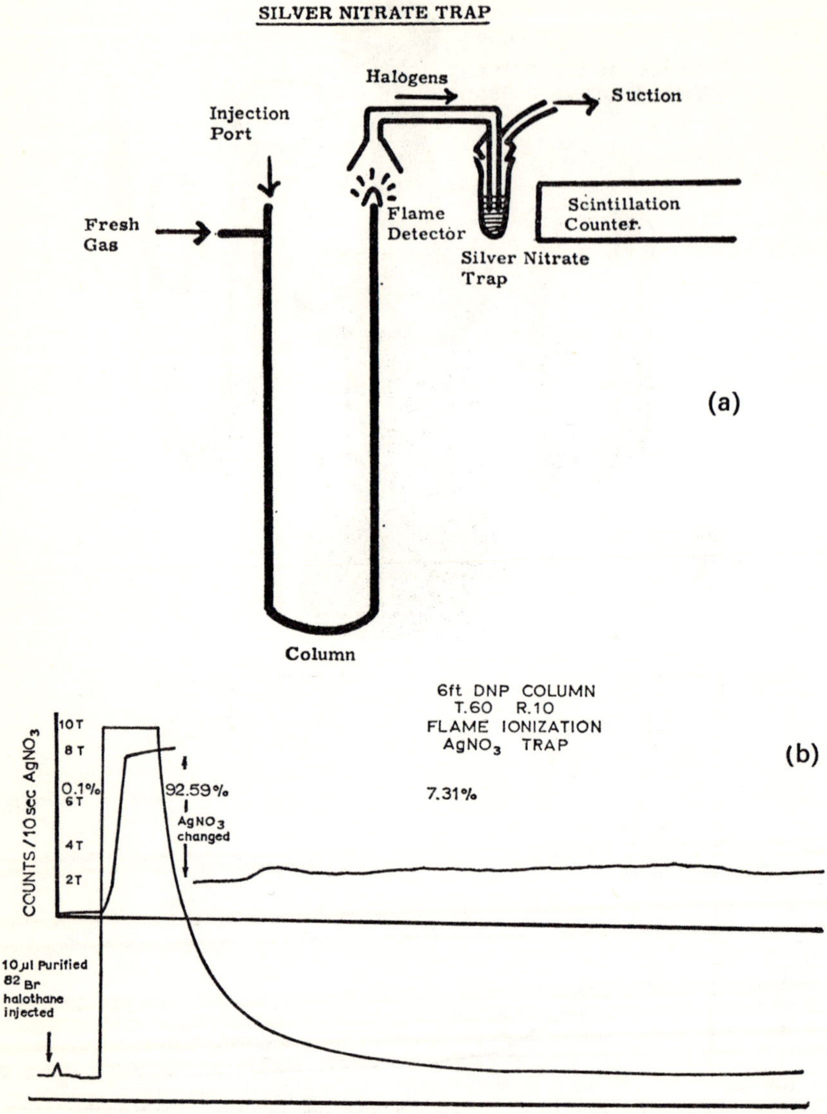

FIG. 3. Method B

(*a*) Silver nitrate trap. Following combustion, halogens were trapped in a solution of silver nitrate (see text).

(*b*) Accumulation of counts showed that this method was more sensitive than method A for detecting [^{82}Br]-containing impurities (0·1% before and 7·310% after halogen peak). T = 1000, thus 10 T = 10,000 counts per 10 sec. (see text).

FIGS. 2 and 3. Following purification by preparative gas chromatography two methods were used to check for the presence of impurities in the same sample of [^{82}Br]halothane.

purified sample in the absence of gas chromatographic peaks radioactivity could be detected, indicating that the material was still impure; (B) halogen compounds were liberated by the combustion of compounds in the hydrogen flame. These were then trapped in silver nitrate which was placed in front of the scintillation detector. Fig. 3*a* shows the use of this method with a further sample of the same partially purified halothane used for Fig. 2. The plot of radioactivity (Fig. 3*b*) shows the localization of [^{82}Br]-contaminants. Comparison of the sensitivity of detection of impurities by these two methods showed that the second method was much more efficient than the first.

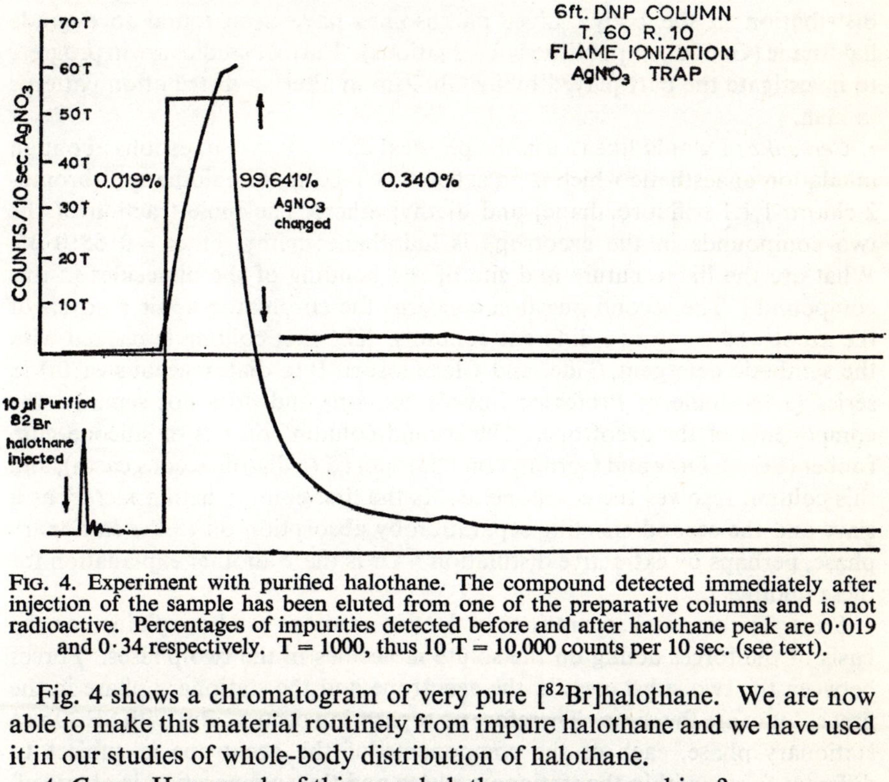

FIG. 4. Experiment with purified halothane. The compound detected immediately after injection of the sample has been eluted from one of the preparative columns and is not radioactive. Percentages of impurities detected before and after halothane peak are 0·019 and 0·34 respectively. T = 1000, thus 10 T = 10,000 counts per 10 sec. (see text).

Fig. 4 shows a chromatogram of very pure [^{82}Br]halothane. We are now able to make this material routinely from impure halothane and we have used it in our studies of whole-body distribution of halothane.

A. Curry: How much of this pure halothane are you making?

Geddes: Our initial sample of 5 ml. of impure irradiated diluted material has to pass through two preparative gas chromatographic columns before it is pure enough for use in man. Only 20 µl. of purified [^{82}Br]halothane was administered by a single breath. We can count radioactivity for 60–90 min. or longer on the low-background counter I have described. We have so far completed a preliminary series of experiments in four individuals who were exposed to radioactive halothane. We are finding residual radioactivity, and theoretical distribution curves are now being plotted and compared with our collected data by our statistician, Dr. Scherrer.

Payne: Metabolic degradation of the halothane may occur during the time you are counting it.

Geddes: Metabolism undoubtedly occurs; the radioactivity counted depends on which part of the curve we look at. There is an uptake curve, associated with [^{82}Br]halothane uptake, then an initial redistribution curve. We do 5-sec. counts at 30-sec. intervals, which allows us to look at the initial uptake and distribution in the body. Liver microsomes have been found to degrade halothane (Geddes, unpublished observations). Further studies are in progress to investigate the part played by metabolism in altering distribution patterns in man.

Cervenko: I would like to ask the physical chemists two questions about an inhalation anaesthetic which is an azeotropic mixture of halothane (2-bromo-2-chloro-1,1,1-trifluoroethane) and diethyl ether. The mole fraction of the two compounds in the azeotrope is halothane:diethyl ether = 0·65:0·35. What are the likely nature and site of the bonding of the molecules in this compound? The second question concerns the chromatographic analysis of the azeotrope using two different columns. The first column is packed with the synthetic detergent, Tide, and Chromosorb P (a diatomaceous earth) in series (a facsimile of Professor Lowe's column) and does not separate the components of the azeotrope. The second column consists of silicone gum rubber (SE 30, Dow and Corning) on Diatoport S (a diatomaceous earth), and this column resolves the components. Is the first column acting merely as a sieve and the second effecting separation by absorption on to the stationary phase, perhaps by extractive distillation? Or is there another explanation for these findings?

Scott: The unequal separations in the two columns can be explained on the basis of the forces acting on the solute molecules in the two phases. Forces between the two substances in the azeotrope and the stationary phase in the first column are the same. Therefore no separation occurs. But with SE 30 as a stationary phase, each of the components of the azeotrope is subject to different forces within the stationary phase and thus a separation is obtained. Concerning your first question, I am not convinced about the nature of the association occurring in the azeotrope. There may not necessarily be chemical bonding at all. Normal intermolecular forces may explain the formation of the azeotrope without our having to assume the existence of any special mechanism such as hydrogen-bonding.

Purnell: Hydrogen-bonding is a most unlikely explanation in a situation where one molecule of oxygen is linked to one of nitrogen. This has never, to my knowledge, been described. But these molecules have very strong dipoles and some sort of charge-transfer complex may explain the linkage. I agree with you, Dr. Cervenko, that it is *extractive* distillation that allows resolution to occur in the second column you described.

Brooks: I would strongly support hydrogen-bonding as the explanation for the link between halothane and ether in this azeotrope.

Lipsky: What is your proof for this?

Brooks: There are two molecules of halothane to each molecule of ether in the azeotrope. The oxygen in the ether has two lone pairs of electrons which are available for hydrogen-bonding. A number of studies based on measurements using i.r. techniques have indicated hydrogen-bonding between chloroform and ether (see for example, Lord, Nolin and Stidham, 1955; Kurosaki, 1962); while dielectric measurements on the chloroform–tetrahydrofuran system have shown two-to-one hydrogen-bonded complexes (Weisbecker, 1965). And chloroform is a close analogue of halothane.

Purnell: Hydrogen-bonding in the ether–chloroform system is not yet unequivocally proven.

Brooks: Additional supporting evidence for such bonding comes from n.m.r. data (for example, Paterson and Cameron, 1963; Bystrov, Lezina and Shostekovskii, 1967) and Raman spectrometry (which uses reflected i.r. rays) (Rezaev and Shchepanyak, 1965). It may be difficult to prove this type of bonding in the ether–chloroform system, but there is no doubt from data from i.r. analyses that ketones (which are not so different from ether) can form two hydrogen bonds with the chloroform molecule (see for example, Whetsel and Kagarise, 1962).

Purnell: There may be an *apparent* analogy about these two systems but it would be difficult to defend your thesis structurally.

Brooks: I see no difficulty. I do not claim that hydrogen–bonding is necessarily a major factor in the gas phase: dipole association may be just as important. But it is certainly feasible that the ether–halothane azeotrope is stabilized by hydrogen-bonding.

Payne: I hope someone will devise an experiment to solve this problem.

REFERENCES

BELLVILLE, J. W., ARTUSIO, J. F., JR., and BULMER, M. W. (1954). *Electroenceph. clin. Neurophysiol.*, **6,** 317–320.

BYSTROV, V. F., LEZINA, V. P., and SHOSTEKOVSKII, S. M. (1967). *Chem. Abstr.*, **67,** 116407*a*.

GEDDES, I. C., MARK, L. C., SCHERRER, J. R., DOMBROWSKI, C., and COHN, S. H. (1969). *Fedn Proc. Fedn Am. Socs exp. Biol.*, **28,** 359.

KIERSEY, D. K., FALCONER, A., and BICKFORD, R. G. (1954). *Anesthesiology*, **15,** 356–364.

KUROSAKI, K. (1962). *J. chem. Soc. Japan (pure chem. sect., Nippon kagaku zasshi)*, **83,** 655.

LORD, R. C., NOLIN, B., and STIDHAM, H. D. (1955). *J. Am. chem. Soc.*, **77,** 1365–1368.

PATERSON, W. G., and CAMERON, D. M. (1963). *Can. J. Chem.*, **41,** 198.

REZAEV, N. I., and SHCHEPANYAK, K. (1965). *Chem. Abstr.*, **63,** 5138*f*.

WEISBECKER, A. (1965). *J. Chim. phys.*, **62,** 1170.

WHETSEL, K. B., and KAGARISE, R. E. (1962). *Spectrochim. Acta*, **18,** 329.

MEASUREMENT OF THE GAS CONTENT OF BLOOD SAMPLES USING GAS CHROMATOGRAPHY

M. J. PURVES

Department of Physiology, University of Bristol

THE principal difficulty which has been encountered in the measurement of the content of gases in blood or other biological fluids by gas chromatography has been in the extraction of gases in a form which is presentable to the column. If the gases are extracted in a modified Van Slyke apparatus, peak resolution is good but extraction is rarely complete (Ramsay, 1959; Lukas and Ayres, 1961; Fahri, Edwards and Homma, 1966). If the gases are washed out of solution by a continuous stream of carrier gas, extraction is virtually complete but peak resolution is poor (Wilson *et al.*, 1961; Wilson, 1964).

A method which combines the virtues of these two previous methods has been developed commercially and by Albers and Fahri (1965*a*, *b*). The principle of the method, the results obtained with it and tests of reproducibility are described.

The apparatus

Fig. 1(*a*) shows the general arrangement of the carrier gas stream in the blood-gas extraction unit and in the standard Beckman gas chromatograph (model GC2A) which is used in conjunction with it. The blood-gas extraction unit is placed upstream of the chromatograph and consists essentially of a stainless steel three-way valve and a reaction tube. By turning the valve to the positions shown diagrammatically in Fig. 1(*b*) the following sequence can be carried out: (1) the dead space of the reaction tube together with gases dissolved in the reagents are purged with carrier gas and the gases are carried to a separate vent; (2) the reaction tube is isolated from the carrier gas stream, the blood sample is introduced and the reaction takes place. Since the tube is full of carrier gas, there exists an effective vacuum with respect to the dissolved gases which then diffuse to equilibrium; and (3) the gases evolved by simple diffusion or after chemical reaction are purged with carrier gas which, at the same time, carries away any residual dissolved gases. These gases are injected downstream into the chromatograph column system.

Two column systems have been used. The first consists of parallel columns and is illustrated in Fig. 1(*a*). Column A consists of 2·4 m. × 6·0 mm. dry fire-brick in series with 1·8 m. × 6·0 mm. Linde 5A molecular sieve. Column B is 0·92 m. × 6·0 mm. activated charcoal.† By substituting the

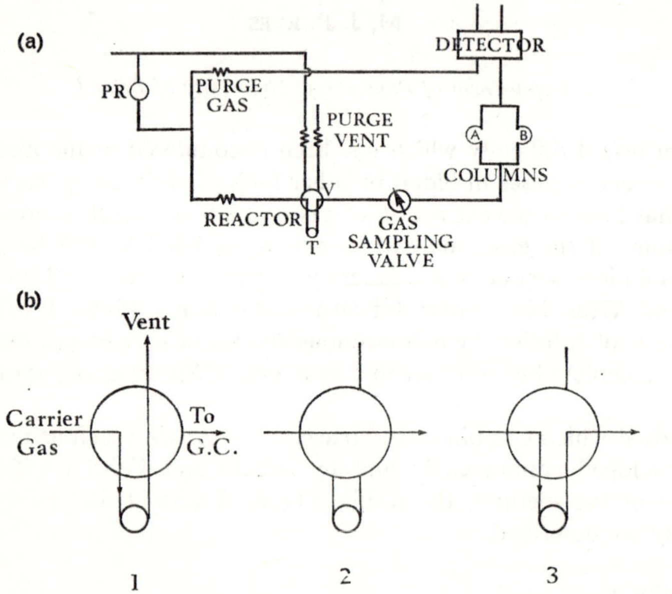

FIG. 1(*a*). Diagram showing the gas circuit in gas chromatograph and blood-gas extraction unit. PR: gas inlet pressure regulator; Reactor V: valve; Reactor T: tube of extraction unit. (*b*) Diagram to show the sequence of positions of the valve in the blood-gas extraction unit. For description of stages of extraction, see text.

latter for the standard silica gel column, the versatility of the instrument is increased, since nitrous oxide can be measured. Further, the operating temperature can be increased from 40° C to 70° C which is the maximum possible with silica gel and the elution times are correspondingly faster. However, with a parallel column system, there is a loss of sensitivity since only a fraction of the gas sample is measured. We have overcome this partially by using a second system in which the charcoal column is eliminated. With such a series column system, carbon dioxide and nitrous oxide are adsorbed

† 4 per cent 2·5 Hexane-dione on Burrells charcoal, Beckman Instruments, Glenrothes, Fife.

in the molecular sieve and cannot therefore be measured. But the sensitivity of the instrument is considerably increased so that the oxygen content of 50 μl. blood samples can be accurately measured and the accuracy of measurement of nitrogen contents of larger samples is improved.

Voltage changes in the thermal conductivity cell are measured with a 1 mV, 1 sec. full scale, 0·305 m. Honeywell recorder set to run at a chart speed of 1·2 mm./min.

Argon-free hydrogen has been used as carrier gas at an input pressure of 15 p.s.i. which gives an effective column flow rate of 90 ml./min. In small laboratories it is probably wise to place a small extractor fan over the instrument. With either type of column arrangement, the thermal conductivity cell is normally operated at 200–230 mA but this can be increased to 300 mA to give greater sensitivity provided the oxygen content of the sample is low since oxygen speeds the deterioration of the filament. With a current of 200 mA and with maximum signal attenuation, the oxygen content of 1·0 ml. of room air gives an approximately full-scale deflection with the parallel column system. Acid-saponin-ferricyanide prepared by the method of Van Slyke and Plazin (1956) is used as reagent, non-ionic polyol as antifoam and N-sodium hydroxide and N-lactic acid to clean the glass reaction tubes.

Procedures

The equipment needs at least 2 hours to warm up and to become stable and it is convenient to leave the instrument switched on but without current to the filaments and with a carrier gas flow of 10 ml./min. overnight before use. Provided that the mains input is adequately stabilized, the baseline remains stable for 24 hours and repeated calibrations agree exactly over a working day.

Gas samples must be introduced precisely at ambient temperature pressure saturated, either through the gas sampling valve or through a carefully calibrated syringe at the liquid sampling valve or reaction tube. The calculated gas volumes are converted to standard temperature pressure dry or, when appropriate, to body temperature pressure saturated.

Blood samples are first drawn into a 1 ml. tuberculin syringe containing heparin and a small washer to facilitate mixing. The syringes are sealed with rubber caps† and the blood transferred via a fine needle to a Hamilton microlitre syringe fitted with 0·1 and 0·2 ml. spacers. The commonest errors in handling derive from the presence of small gas bubbles and leaks

† 1 ml. vial turnover closures, obtainable from William Freeman, Suba-Seal Works, Staincross, Barnsley, Yorkshire.

developing around the needle as the blood sample is injected through the rubber cap into the side arm of the reaction tube.

Calibration

In a preliminary series it was established that the relation between peak height and concentration of oxygen, carbon dioxide and nitrous oxide was linear, passing through the origin. For these gases, therefore, the instrument has been calibrated each day with two volumes (0·25–1·0 ml.) of room air for oxygen, 0·5 ml. of two carbon dioxide in air mixtures (3·0 and 6·0 per cent carbon dioxide) whose composition was accurately determined for other purposes with a Lloyd-Haldane apparatus and two volumes of pure (99·23 per cent) nitrous oxide, injected by syringe into the liquid sampling valve or reaction tube. For carbon monoxide calibration the same procedure is followed, using two volumes of pure (99·65 per cent) carbon monoxide. However, the relation between peak height and carbon monoxide concentration is not linear, the carbon monoxide peak being asymmetric with variable base. It is therefore necessary to measure the area of the peak using either planimetry, millimetre graph paper or, better, with an integrator. To make the calibration independent of other methods of analysis, the instrument can be calibrated for carbon dioxide by preparing bicarbonate solutions in concentrations which bracket the expected range of carbon dioxide. A correction must be made with respect to the oxygen peak since argon in the atmosphere (0·98 per cent) contributes to the oxygen peak. The correction factor of 0·048 obtained by Lukas and Ayres (1961) has been used. Since argon is present in blood only in the dissolved form, no correction for blood argon is necessary.

Results

Figs. 2 and 3 show representative chromatograms using the parallel column arrangement, and demonstrate the size and sequence of peaks and elution times. An important point to note in Fig. 2 is that whereas the nitrogen contained in 0·1 ml. of room air (approximately 80 µl.) gives rise to an easily measurable peak, the peak from nitrogen dissolved in 0·1 ml. blood is barely perceptible. Although extraction of nitrogen has not been studied in detail, it has been shown that with the parallel column arrangement, where the volume of nitrogen is 50 µl. or above, satisfactory peaks are obtained and the coefficient of variation (C.V. = S.D./mean value) varies between 0·27 and 1·34 per cent. But even when 0·2 ml. blood samples are used, peak heights of 9–12 mm. are rarely exceeded and the C.V. rises to ± 3·75 per cent. With the series column system, nitrogen in 0·1 ml. blood gives easily measurable peak heights of 18–26 mm. and in one series of eight repeated estimations the coefficient of variation was 0·67 per cent. Fig. 3 illustrates the additional peaks obtained with carbon monoxide and nitrous oxide with their elution times. Fig. 4 illustrates the chromatograms obtained with 50 μl. air and blood and the increased sensitivity achieved by using a single fire-brick and molecular sieve column.

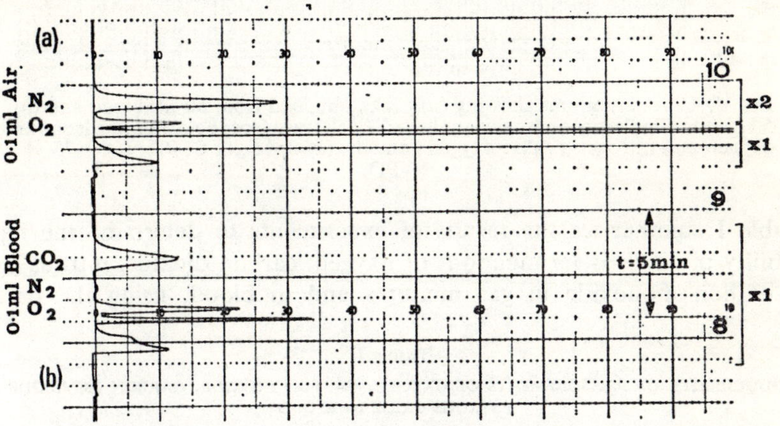

FIG. 2. Chromatograms showing peaks obtained from (*a*) 0·1 ml. room air and (*b*) 0·1 ml. blood. Sequence of peaks, from below upward, and volumes of gases represented (μl., standard temperature pressure dry (STPD), corrected for argon) are (*a*) Air: pressure artefact; composite peak for all gases in sample; oxygen = 20·01; (*b*) Blood: oxygen = 16·4; carbon dioxide = 24·5. *t* = 5 min.

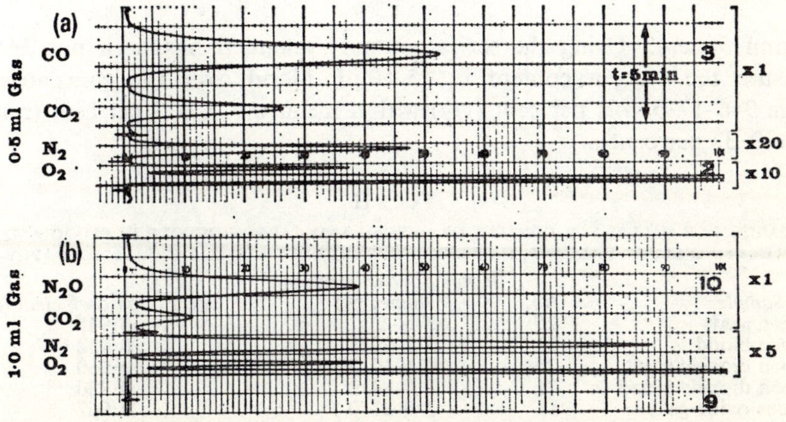

FIG. 3. Chromatograms to show additional peaks of, (*a*) carbon monoxide (81 μl., STPD, gas sample size = 0·5 ml.) and (*b*) nitrous oxide (108·5 μl., STPD, gas sample size = 1·02 ml.). *t* = 5 min.

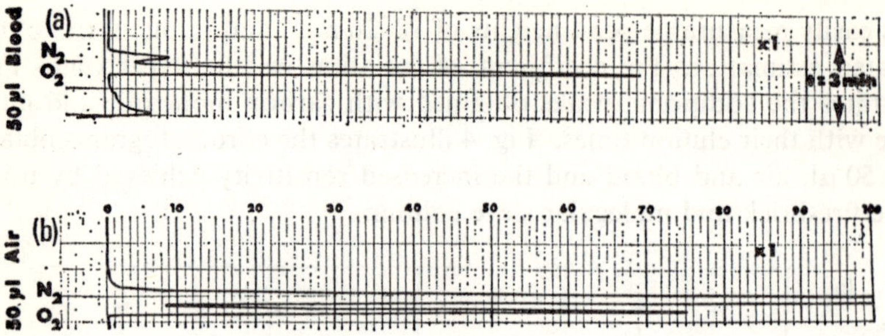

FIG. 4. Chromatograms showing (*a*) peaks obtained from 50 µl. blood and (*b*) 50 µl. air. Single molecular sieve column. Each trace from below upwards shows oxygen and nitrogen peaks: oxygen content: (*a*) 8·44 µl., STPD; (*b*) 9·75 µl., STPD.

Table I summarizes the results of experiments to determine the reproducibility of single determinations of oxygen, carbon dioxide, nitrous oxide and carbon monoxide in gas mixtures and in blood, using the parallel column system. Using the series column system, it has been possible to measure the oxygen content in 25–50 µl. blood over the concentration range 9·6–17·5 vols per cent (v/v) with a mean C.V. ±0·3 per cent (range: 0·21–0·37 per cent).

TABLE I

REPRODUCIBILITY OF MEASUREMENTS OF OXYGEN, CARBON DIOXIDE, CARBON MONOXIDE AND NITROUS OXIDE IN BLOOD

Sample	*Volume (ml.)*	*No.*	*Range of concentration (volumes %)*	*Mean coefficient of variation (%)*	*Range of coefficient of variation (%)*
Oxygen blood	0·1	23	3·5–19	±0·27	0·14–1·10
Carbon dioxide blood	0·1	14	11·0–56	±0·32	0·11–0·64
Nitrous oxide blood	0·2	11	11·0–35·0	±0·41	0·23–0·93
Carbon monoxide blood	0·2	11	11·5–33	±0·98	0·67–1·28

TABLE II

COMPARISON OF VALUES FOR CONTENT OF OXYGEN AND CARBON DIOXIDE IN BLOOD AND GAS SAMPLES USING LLOYD-HALDANE, NATELSON-KOPP METHODS AND GAS CHROMATOGRAPHY

Sample	*No.*	*Range of concentration (volumes %)*	*Correlation coefficient (r)*
Oxygen gas†	25	8·5–23	0·945
Oxygen blood	21	3·2–20·8	0·932
Carbon dioxide gas†	25	0·6–10·2	0·960
Carbon dioxide blood	21	1·0–84·0	0·920
Nitrous oxide gas‡	11	3·0–27·0	0·967

† Chromatograph peaks for oxygen corrected for argon.

‡ Nitrous oxide content of nitrous oxide/oxygen mixtures measured in Lloyd-Haldane apparatus by difference after measurement of oxygen.

Table II gives the results of a series of studies in which samples of blood and gas were analysed in parallel for their content of oxygen, carbon dioxide and nitrous oxide using a Lloyd-Haldane and Natelson-Kopp apparatus and gas chromatography.

Discussion

This method of extracting gases from blood and measuring them with gas chromatography appears to have a number of advantages over the traditional methods. First, the semi-automatic method of extraction means that with comparatively little fatigue on the part of the operator a large series of analyses can be carried out as an experiment proceeds. Second, the method makes it possible to analyse the content of a number of gases in each blood sample. Traditionally a comparatively large sample would be needed and, in addition to a Van Slyke or Haldane apparatus, infra-red (i.r.) detectors for carbon monoxide and nitrous oxide would be necessary while the methods of extraction are cumbersome and time consuming (Lawther and Bates, 1953; Coburn *et al.*, 1964). Third, the method enables small air leaks to be detected.

When equivalent volumes of blood or gas are analysed, it is clear from Tables I and II that gas chromatography matches the Haldane and Van Slyke methods in accuracy with respect to oxygen and carbon dioxide. The C.V. was greater for blood than that reported by Lawther and Bates (1953) who used i.r. spectroscopy but it is possible that the accuracy was equivalent since these workers used 5 ml. blood samples. The accuracy of carbon monoxide determination was considerably less than the accuracy for other gases and this may have been due to the graphical method of measuring the peak area: it could probably be improved by the use of an integrator. Even so, the mean C.V. obtained over similar ranges compares favourably with that reported by Roughton and Root (1945), who used a vacuum syringe method, Coburn and co-workers (1964) who used i.r. spectroscopy and Ayres, Criscitiello and Giannelli (1966) who used a modified Van Slyke apparatus and gas chromatography.

The sensitivity of the method can be increased by using a single column system (see Fig. 4) so that 50 μl. blood samples can be analysed with considerable accuracy. The use of this column system, however, restricts the variety of gases that can be measured. The method would seem to be suitable for routine laboratory use. In addition, there are special circumstances, for example the analysis of very small blood samples or those with low carbon dioxide content from reptiles and fish, in which the accuracy of the method of extraction and measurement with gas chromatography cannot be matched.

SUMMARY

Current methods of measuring the content of respiratory and other gases in blood using gas chromatography are described. The accuracy of the measurement of these gases in 50/20 ml. blood samples is at least as good as the more traditional methods of analysis, the techniques involved are simpler and the time consumed is shorter. Examples of the usefulness of the method are given and ways in which the sensitivity of the instrument can be increased are discussed.

REFERENCES

ALBERS, C., and FAHRI, L. E. (1965*a*). In *Lectures in Gas Chromatography*, vol. 2, pp. 163–187, ed. Mattick, L. R., and Szymanski, H. A. New York: Plenum Press.

ALBERS, C., and FAHRI, L. E. (1965*b*). *Z. ges. exp. Med.*, **139**, 485–505.

AYRES, S. M., CRISCITIELLO, A., and GIANNELLI, S. (1966). *J. appl. Physiol.*, **21**, 1368–1370.

COBURN, R. F., DANIELSON, G. D., BLAKEMORE, W. S., and FORSTER, R. E. (1964). *J. appl. Physiol.*, **19**, 510–515.

FAHRI, L. E., EDWARDS, A. W. T., and HOMMA, T. (1966). *J. appl. Physiol.*, **18**, 97–106.

LAWTHER, P. J., and BATES, D. V. (1953). *Clin. Sci.*, **12**, 91–95.

LUKAS, D. S., and AYRES, S. M. (1961). *J. appl. Physiol.*, **16**, 371–374.

RAMSAY, L. H. (1959). *Science*, **129**, 900–901.

ROUGHTON, F. J. W., and ROOT, W. S. (1945). *J. biol. Chem.*, **160**, 123–133.

VAN SLYKE, D. D., and PLAZIN, M. (1956). In *Quantitative Clinical Chemistry. (Methods)*, vol. 2, p. 321, ed. Peters, J. P., and Van Slyke, D. D. Baltimore: Williams and Williams.

WILSON, R. H. (1964). *J. Gas Chromat.*, **2**, 365–373.

WILSON, R. H., JAY, B., DOTY, V., PINGREE, H., and HIGGINS, E. (1961). *J. appl. Physiol.*, **16**, 374–377.

DISCUSSION

Hill: Until 1967 we used a similar apparatus to the Beckman machine you described. But we had to stop using it although it worked well because the manufacturers stopped producing the sintered Teflon frits that were a vital part of the blood-gas extraction system. In the UK it is impossible to buy a frit of suitable pore size; if the pores are too small the gas will not pass through them and if they are too large the blood leaks out. These frits are only available in the US. It took us about a year to develop an alternative system for detecting blood gases.

Robinson: We had this machine too, but it became unusable because of the difficulty in obtaining the Teflon frits and because of leakage of blood.

Carter: Why was this instrument withdrawn when sales were so high?

Did the equipment not tackle the right problem or was it poorly designed for the job in question?

Purves: We were told only that sales figures for this equipment were unsatisfactory: these may have been the sales in the US, not in this country. But it seemed an extraordinary decision when it was clear from the beginning that this method could replace all previous traditional instruments for blood-gas analysis. Perhaps those concerned with blood-gas analysis may reckon it is cheaper to pay technicians to do the analyses with the old machines than to spend money on a new gas chromatograph.

Carter: Was it basically a question of cost or of technical specifications?

Purves: I should call it a question of *misjudged* cost.

Payne: Most of the groups who use a gas chromatograph for blood-gas analyses in the UK have designed their own extraction device. The industry has ignored this aspect of the problem and concentrated on providing equipment to do the analysis *after* extraction. Most of us make these machines work because we have designed the extraction component ourselves.

Hill: Is a helium ionization detector a satisfactory device for analysing blood gases?

Lipsky: We have tried various ionization processes to detect permanent gases (not in blood, incidentally, but in calibrated mixtures). We found that if one sets up an ionization detector and measures the drift velocity of ions within it, one can indeed measure quite small quantities of gases. We obtained a full-scale deflexion for all components from a sample volume of one millilitre of air. The main problem with this detector was the limited linear dynamic range—it would only extend to a range between a few hundreds and one. The helium ionization detector is useful for trace analysis but can be too easily saturated.

Scott: Oxygen, carbon dioxide, carbon monoxide and nitrous oxide can all be effectively detected using the Cropper conversion system (Cropper, Heinekey and Westwell, 1967) and flame ionization detector in conjunction with a molecular multiplier. To detect the carbon dioxide or oxygen the sample would be passed over carbon, and the carbon monoxide produced converted quantitatively to methane. The unit is conveniently small and can be easily fitted into the existing chromatograph. Using this system, a flame ionization detector would give overall sensitivity for oxygen of about 1 ng./ml. The system would not detect nitrogen unless one could devise a way for nitrogen to release another compound that contains hydrogen and oxygen.

Purves: Nitrogen tends to be neglected; we badly need a technique for measuring it. Although nitrogen is often dismissed as a tiny "blip" on the chromatogram, its presence in alveolar gas and in blood is of considerable biological importance (see for example, Rahn, 1961). Chromatography is the only satisfactory technique I know by which nitrogen in blood can be reliably measured.

Payne: Accurate quantitative measurements of nitrogen are essential for the study of pulmonary function. Have you any comments on this, Professor Robinson?

Robinson: The Holy Grail for the respiratory physician is the measurement of the partial pressure of nitrogen in the blood. The pulmonary defects causing hypoxia and cyanosis clinically can be accurately defined from measurement of the three blood gases—oxygen, carbon dioxide and nitrogen. Surprisingly, nitrogen is not in equilibrium with air. One way of measuring blood nitrogen is by the estimation of urinary nitrogen. It is relatively easy to measure nitrogen in the urine just because urinary volume is reasonably large, but it is not known if nitrogen in urine and blood are in equilibrium. With the method that has been used by Dr. Hill and his colleagues (Denison Davies, 1969) we were able to measure blood nitrogen partial pressures from 50 μl. of blood with a reproducibility of ± 2 mm. Hg; but unfortunately this technique is only reliable in experienced hands.

Scott: This confirms my support for an instrument that utilizes the flame ionization detector. Such an instrument provides a sensitivity one or two orders greater than any other method and does not need highly skilled human technicians to operate it. Another possible way of overcoming the difficulties with nitrogen would be to pass it over sodium hydride. Would this release hydrogen?

Martin: No; not hydrogen but solid sodamide would be formed. But if nitrogen were passed over calcium hydride, calcium nitride would be produced and this would release an appropriate amount of hydrogen. That might be a better way of doing it.

Robinson: We would still have to deal with the problem of handling the blood and getting it into the shaker.

Scott and Martin: You just put it on the end of the column.

Robinson: We have to do our estimations in an atmosphere of helium because the volume of nitrogen in the blood is so small; by the time it is injected into the shaker there is an air peak, and this ruins the estimation.

Scott: You are now talking about the sensitive detection of nitrogen—a different question from the one we were discussing.

Hill: The use of a molecular multiplier with a flame ionization detector sounds promising. Although we ourselves are not concerned with the quantitative estimation of nitrogen we have found the nitrogen peak to be a useful index of air leaks. So, for a rather different reason from yours,

Professor Robinson, we try to keep the nitrogen peak as small as possible.

Robinson: If a reliable technique for measuring blood nitrogen were available the manufacturers would be forced to develop better methods for handling the blood samples. Most respiratory physiologists would jump at this.

Purves: It would be extremely useful if we could combine the molecular multiplier system with Dr. Janák's technique of fractional analysis. The respiratory system, like most control systems, tends to be cyclic. Thus oxygen and probably carbon dioxide levels in arterial blood show cyclic changes which have the same period as respiration. These cyclic changes may perform an important function in the control of respiration. We need a technique to measure the gas *content* of the blood, rather similar to the one that Dr. Janák described for the respiratory gases in expired air, although the latter situation creates fewer problems.

Martin: Why is it difficult to determine these cyclic changes?

Purves: The difficulty is not connected with the chromatography but with the handling of the blood so that a continuous or a rapid fractional analysis can be made.

Martin: It would be quite easy to fill a long capillary tube with blood and divide the tube at frequent intervals. If the tube were fitted direct to the needle and the time for filling were several cycles, wouldn't this solve the problem?

Purves: This would be all right if one could avoid turbulence and, thus, mixing. But some degree of mixing is likely to occur in a rigid tube.

Martin: Not if the tube is narrow enough.

Purnell: Is part of the problem that there is not enough pressure to push the liquid along?

Purves: Yes, because this limits the cross-section and length of the tube.

Purnell: You could just put a tube into a blood vessel and sample the low-volume, continuous stream intermittently.

Purves: Cyclic fluctuations of the partial pressure of gases in blood can be measured. It is also important to measure the corresponding fluctuations of gas concentrations in expired air and I hoped that combining Dr. Janák's technique and the molecular multiplier system would help us to do this.

Martin: How large are these fluctuations?

Purves: They are of the order of a partial pressure of 6 mm. Hg—say a pressure of 100 mm. Hg, $\pm$ 3 mm. Hg. The fluctuation of gas concentrations depends on various factors, of course.

Purnell: What is meant by expressing the tension of a gas in blood in terms of millimetres of mercury? Is this relative to a gas-phase system?

X mm. of gas in blood is a curious way of expressing the concentration of a gas in a liquid.

Purves: We are expressing not the concentration but the pressure that a gas exerts in the liquid phase: for example, oxygen held in physical solution in plasma. The pressure will be determined by equilibrium with gas in the gas phase.

Purnell: In other words "tension" is the pressure that a given number of "millimetres tension in solution" would exert in the gas phase—a factor, essentially, with the usual properties of a partition coefficient.

Purves: Yes.

Scott: Why not express these gases as their absolute concentrations in blood?

Purves: This is often done in practice and these concentrations can be measured directly. But difficulties arise when blood-gas content is expressed as a concentration from measurement of partial pressure because this involves non-linear dissociation curves, and the position of these curves, also, is affected by a number of factors.

Purnell: Your discussion of partial pressures makes the problem of measuring blood nitrogen sound more difficult than it probably is in practice. The notion of blood tensions is confusing to people like me who are not used to it.

Payne: The whole concept is confusing to everyone!

Robinson: We think of the blood gases in this way because it is the partial pressures that govern their distribution in lung, blood vessels, brain and other organs.

Purnell: The partition coefficients in all circumstances must be known or these measurements are meaningless.

Martin: In tracing the movement of gases from one cell to another is one interested primarily in the pressures?

Purves: Yes.

Scott: If one used the haemoglobin concentration or oxygen content of blood, this same value of 6 mm. Hg would mean something quite different.

Purves: Absolutely. We are talking about millimetres of mercury but obviously the fluctuation in *absolute concentrations* will vary enormously over the particular range.

Payne: If haemoglobin is removed from a blood sample, although the tension of oxygen may be unchanged the amount of available oxygen is substantially reduced.

Hill: A useful characteristic of the gas chromatography technique is that it forces one to think in terms of *content*.

Payne: It also most usefully forces us to reconsider the oxygen dissociation curve. Standard curves were originally formulated from rather narrowly based studies on a few individuals only (Haldane, 1922 [Haldane and Priestley, 1935]; Bock, Field and Adair, 1924).

Hill: If one is studying dissociation curves, which in experimental circumstances may not be the same as the standard curves in the physiology textbooks, speed is essential. In the laboratory, gas contents in 30 blood samples and 30 controls can be quickly determined with a gas chromatographic system (Hill, 1966). One would be daunted by the time taken for such estimations using the Van Slyke apparatus.

Lipsky: A system of molecular multipliers combined with a flame detector sounds a most promising way of analysing the respiratory gases, especially if a technique can be found for measuring nitrogen. So much is already known about the flame detector, and it has such a wide linear dynamic range and sensitivity, that this combination should be excellent.

Hill: We are all used to looking at organic materials in smaller and smaller quantities, but many technical difficulties persist in our analyses of the blood gases. A sensitive technique for detecting them would be invaluable.

Martin: The helium detector should be workable for all these gases, including nitrogen.

Robinson: Gas chromatography has two great advantages in the measurement of blood gases; first, the technique is idiot-proof. The methods currently in use for these measurements depend absolutely on the mood of the technician—gas chromatography does not; and, second, gas chromatography measures, and so makes us think in terms of, oxygen *content*. What matters is how much oxygen the heart pumps out into the blood every minute (that is, content, which is nothing to do with partial pressure). But no one seems prepared to develop gas chromatographic methods for blood gas estimations commercially. So much thought, work, time and money go into the development of equipment for the analysis of blood gases in clinical work that surely, if someone really worked at the problem, better techniques could be developed.

Hill: Many quite satisfactory commercial devices for estimating blood pH, and carbon dioxide and oxygen tensions, are available and used regularly in hospitals in the UK. This is a fairly consistent group of apparatus, and technicians can be trained to use them without too much difficulty. In contrast, a chromatograph estimates content. Physiologically, both gas content and gas tensions are needed. The trend has been for clinicians to think in terms of gas tensions and pH and to view gas chromatography and gas content as the province of the chemist.

Payne: There are trends in medicine as in everything else. Until we can persuade our clinical colleagues to look at content, the pressure on industry to produce the necessary equipment will not exist.

Lipsky: If the clinicians in this group could reach agreement about a satisfactory sampling system for the gas chromatograph, then surely the physical chemists here could work out satisfactory and appropriate instrumentation. The technology to do this is already available. We do not need a breakthrough but we do need communication between clinicians and basic scientists.

Hill: The sampling end of the chromatograph and the extraction systems work well in most hands now. But we still need ways of increasing detector sensitivity.

Lipsky: This is not the whole story. Although with the chromatograph we can measure gases in a blood sample of one microlitre we will not always be satisfied with this. Fifty millilitre used to be the standard sample size, we now have a system sensitive enough to work on one microlitre, and increasing sensitivity will be constantly in demand.

Hill: I agree. Using a molecular multiplier with a chromatograph is an improvement, but when this combination is incorporated in our standard equipment, the extraction apparatus now in use may no longer be adequate. And so on.

Martin: Is anyone interested in measuring argon?

Robinson: Some of the members of my department have used argon. Because it is an inert gas, much can be learned about respiratory function when it is used as a tracer (Jones and Clarke, 1969).

Martin: Are you using argon as distinct from nitrogen?

Robinson: Both are useful. Reliable methods for measuring the blood content of inert gases like nitrogen and argon would open up areas of respiratory physiology that have never yet been studied or can only be studied with radioactive gases.

Hill: Gas chromatography is the poor man's mass spectrometry. The few scientists with mass spectrometers can use argon in respiratory physiological studies, but the majority of us have to make do with the gas chromatograph for economic reasons.

Blackmore: A problem in extracting the gases from blood before gas chromatographic analysis concerns sample size. When blood is heated to a temperature of 180° C carbon monoxide is liberated quantitatively. But one needs a sample of 50–100 μl. of blood for the liberation of a measurable amount of carbon monoxide. If one could use a very small volume—1 μl.—, and liberate the gases through heating, then no extraction would be necessary.

Scott: If carbon monoxide is passed directly through the Cropper nickel catalyst (*loc. cit.*) and then into a flame ionization detector, carbon monoxide can be detected at the same low levels at which organic vapours are detected by the flame ionization detector.

Blackmore: That is a very exciting idea.

Robinson: Incidentally, when blood is heated nitrogen is released from plasma proteins. This may falsify the estimation because this nitrogen is not what we want to measure.

How did you heat the column, Mr. Blackmore, and was it a pre-column?

Blackmore: I heated a glass tube containing molecular sieve SA to a temperature of 180°C with an electric Bunsen burner, and injected the sample into one end of the pre-column. All the gases were liberated immediately and were transferred to a gas chromatograph via a gas loop. The water emerged from the pre-column well after the permanent gases and could be bypassed into the atmosphere thereby preventing column contamination.

Lipsky: Could you sample repeatedly?

Blackmore: This was not a problem because I was looking for carbon monoxide only. The gadget I made could not deal with more than ten samplings, and ceased to work properly if I introduced 1 ml. or more of water. But if one could increase sensitivity by using a flame ionization detector, and small volumes (1 μl. or less) of sample, this system could be very useful. Another advantage is that it does not need alkaline ferricyanide to liberate the blood gases.

Lowe: I have done some preliminary studies on a negative peak type of chromatography. We use cyclopropane as a tracer, in nitrogen as the carrier gas, in a hydrogen flame detector. This mixture is passed over a platinized asbestos column at a temperature of about 400°C. When 1 μl. of blood is injected into this system a negative peak is recorded due to oxidation of the trace of cyclopropane. This is a very sensitive method for detecting oxygen.

Scott: If one uses a negative signal in the flame-ionization-detector system the signal range is from $0\text{–}10^{-12}$ A, which is too small to accommodate the range of concentrations needed. But if one operates detectors such that positive ionization occurs and a positive signal is obtained, sensitivities as small as 1 pg./ml. can be realized.

Lowe: Using negative peaks at one amplifier sensitivity setting, the instrument can give direct readings for oxygen contents of 0–25 per cent (by volume).

REFERENCES

BOCK, A. V., FIELD, H. JR., and ADAIR, G. S. (1924). *J. biol. Chem.*, **59**, 353.
CROPPER, F. R., HEINEKEY, D. M., and WESTWELL, A. (1967). *Analyst, Lond.*, **92**, 443–449.
DENISON DAVIES, D. (1969). *Br. J. Anaesth.*, **41**, 192.
HALDANE, J. S. (1922). *Respiration.* New Haven: Yale University Press. (Haldane, J. S., and Priestley, J. G. (1935). *Respiration.* London: Oxford University Press.)
HILL, D. W. (1966). In *Oxygen Measurements in Blood and Tissues*, pp. 63–80, ed. Payne, J. P., and Hill, D. W. London: Churchill.
JONES, J. G., and CLARKE, S. W. (1969). *Clin. Sci.*, in press.
RAHN, H. (1961). *Harvey Lect.*, **55**, 173–199.

GENERAL DISCUSSION

Hill: Could one use an electrolytic generator with the hydrogen flame detector? I have never seen this combination of apparatus. Are there any practical problems in generating the required flow to the flame detector? It does not need much current does it, Dr. Scott?

Scott: I have not yet used this system myself, mainly because gases are piped round our laboratories, but such a generator has been used by other workers at Unilever and has been found to work well and convincingly for long periods of time. The generator needs recharging every month or so but this is hardly a problem.

A. Curry: What pressure is needed?

Scott: A hydrogen and oxygen flow of 6·9–13·8 kN/m.^{2} (1–2lb./in.^{2}) for most flame ionization detectors, but column inlet pressures can be up to at least 690 kN/m.^{2} which means that for column flows the system might be inadequate.

Carter: A commercially available instrument that combines an electrolytic generator and a hydrogen flame detector, with a diffusion cathode, can quite easily generate a pressure of 345 kN/m.^{2}.

Purnell: A wrong connexion on an electrolytic generator can be disastrous.

Lipsky: Have any accidents occurred with the hydrogen cylinder?

Purnell: Confusion exists about the explosive qualities of hydrogen. Hydrogen is far less dangerous than is generally supposed. Hydrogen–oxygen mixtures explode only within specific pressure limits. The real danger with hydrogen is fire not explosion; and a fire risk is relatively easily handled. It is odd that the much less important explosion risk is what bothers people.

Carter: I have seen an explosion when a hydrogen cylinder was connected to the air line; the oven filled with hydrogen and exploded when the flame was lit. But this was an explosion of the chromatograph oven, not the gas cylinder.

Scott: The safety officer at British Oxygen Manufacturers recently told me that they had never known a hydrogen cylinder to cause trouble by explosion in a laboratory.

Carter: The problem concerns the plumbing to the chromatograph not the hydrogen cylinder. If the same size of tube is used throughout tubes may be wrongly connected, or if plastic tubing is used it may melt.

Scott: The same risk applies to other gas cylinders, calor gas and so on.

Carter: Of course. Generators produce a pressure of 345 kN/m.^{2} of hydrogen but a much lower pressure of oxygen, which makes them less dangerous.

Martin: One can easily make hydrogen by electrolysing a solution of hydrogen bromide; the bromine sinks to the bottom of the vessel and only hydrogen is released.

Scott: Safety devices can be incorporated into these apparatus.

Purnell: This equipment should be an off-the-shelf item like a dry battery. The incorporation of safety devices and so on turns it into a research project. I see no problems with the hydrogen cylinder.

Scott: Another advantage is that light-weight hydrogen cylinders are now available.

Sjövall: The strong alkaline solution in an electrolytic generator is another danger. If the instrument is being carried round the alkali can easily spill or get to the detector via the tubing. This has happened to me. I, too, prefer to use a hydrogen cylinder.

Gray: No one considers that the mobile phase in gas chromatography can be anything but an inert gas, used as an inert carrier. One of the flexibilities of liquid–solid or liquid–liquid chromatography is that the eluting solvents can be altered at any stage during the run; in other words, chromatography can be controlled while it is taking place. Could one do the same sort of thing with a gas chromatographic system? If one wished to separate two close-running compounds, say two saturated lipids, could one introduce into the gas flow a controlled concentration of another solvent—chloroform vapour for example—thereby altering the partition coefficients of the two lipids with respect to the mobile and stationary phases, and thus possibly enhancing the degree of separation? Introduction of a vapour into the mobile phase would increase the detector baseline but this could be stabilized at the beginning of the experiment. Professor Martin, do you think that such a system is feasible?

Martin: No. The extent of the change to be expected in the activity in the gas phase would be very small compared to the change that the same substance would make in the solvent of the stationary phase. Whatever was introduced would tend to change the stationary more than the gas phase. The only scientists that I know of who have attempted this sort of thing are Sie and Rijnders (1967). Rijnders has worked with the gas close to its critical point (an uncomfortable situation), at which the partition is drastically altered in favour of the gas phase.

Lipsky: What Rijnders has done with the dense gas phase we can do equally well in the liquid phase. Incidentally, the potential hazards of using gases at such very high pressures in the laboratory are very great.

Martin: I agree that this is not a very practical development.

Lipsky: Steam has been used to saturate certain very polar compounds as a means of separating them.

Martin and Scott: But this also exerts its effect on the stationary phase.

Scott: Interactions in the mobile phase, if this is a vapour or gas, at normal pressures are very small but increase as the pressure increases.

REFERENCES

SIE, S. J., and RIJNDERS, G. W. A. (1967). *Analytica chim. Acta*, **38**, 31–44.

RECENT DEVELOPMENTS IN THE USE OF GAS CHROMATOGRAPHY IN FORENSIC TOXICOLOGY

A. S. CURRY

Home Office Central Research Establishment, Aldermaston, Berkshire

THE forensic toxicologist is concerned with analyses from both the living and the dead and he attempts to discover the presence of any toxicologically significant substance that may have contributed to the illness or death of the patient. Within the compass of this definition comes a wide variety of areas of interest—blood and urine alcohol determinations in relation to the Road Safety Act,† investigations into suspected transgressions of the Drugs (Prevention of Misuse) Act or the Dangerous Drugs Act and Regulations, malicious administration of poisons or noxious substances with intent to murder, injure, aggrieve or annoy, or perhaps with intent to procure an abortion, and, finally, the investigation of sudden deaths of all categories to assist police and coroners' inquiries.

It is of historical interest that the first use of gas chromatography in forensic toxicology was in 1955 with the comparison of a cigarette-lighter fuel with a sample extracted from a bottle of medicine into which the fuel had been put maliciously (Curry, 1955). The increase in the use of gas chromatography in forensic science since this time has been dramatic, and all the Home Office Forensic Science Laboratories now use the technique extensively. It is worth stressing, as one of the "recent developments" that form the subject of this paper, that in forensic toxicology the use of gas chromatography has shown the same exponential rise as has been noted for the use of this technique in other areas of science. The range of its use stretches from analyses for gases—of which carbon monoxide is the most important—, through vapours—following the inhalation of petrol for example—, to liquids themselves—of which alcohol is the most common—and virtually all the pharmaceutical armoury, whether the drugs concerned are barbiturates, psychotropic drugs, narcotics or hallucinogens. The very wide range of compounds that are amenable to gas chromatographic separation has brought into the forensic science laboratory problems that are probably not met to such an extent in other areas of interest. It is not possible to equip our

† Legislation refers to the UK.

laboratories with such lavishness that machines can be kept for special purposes—one for amphetamines, one for gases and so on—although, because of the nature of the work in forensic science, these are all recurring problems and much time is wasted in changing columns, especially when it is necessary to break into routine analyses to answer an urgent police request. Perhaps by defining the problem we may encourage manufacturers to produce machines that are capable of running, say, four columns in parallel with four gas-flow systems and four separate detectors. If provision for different ranges of temperature can be made, so much the better, but I stress the need for increasing the number of available independent columns without increasing the space required for the ovens and gas cylinders.

Coupled with the need for greater flexibility in column changing is the need for easy changing of detectors; in the course of the analyses of one body a katharometer for carbon monoxide, a flame ionization detector for alcohol, barbiturates and so on and an electron capture detector for organo-chloro pesticides, anaesthetics and so forth may all be needed. The toxicologist could also make good use of specific nitrogen and phosphorus detectors in his general search for poisons. This need for flexibility in working conditions is the forensic toxicologist's outstanding need.

The value of gas chromatography to the forensic scientist is not only its power of resolution but also its sensitivity; through its use evidence of significant traces of toxic compounds, previously undetectable, can be found. If a woman is semi-anaesthetized by a "rag soaked in a smelly liquid" before being raped, the analysis of a few millilitres of her blood to detect ether, chloroform or even carbon tetrachloride is now routine and invaluable to the police inquiry. Similarly, it is possible to analyse tissues that, before the advent of gas chromatography, were useless—heart muscle for carbon monoxide for example—in cases of burning, in which it may be vital to know if the victim was alive when the fire started.

The use of gas chromatography in forensic toxicology in the UK deserves a special mention as far as alcohol is concerned; in 1965, when the new legislation on road safety proposed to make it an offence to drive a motor vehicle with a blood alcohol concentration exceeding 80 mg./100 ml., the British Medical Association had recommended the use of capillary blood for assessing the amount of alcohol in the body. The task of analysing 10 μl. of blood for alcohol at this concentration, that is, accurate quantitative measurement of 8 μg. of alcohol with a hoped-for standard deviation of less than 2 per cent, was given to Home Office scientists. Gas chromatography was the obvious technique to use and a recent study of the method used by the Forensic Science Laboratories in the UK (this method involves dilution with an internal standard, gas chromatographic separation and integration) has shown precision well below a standard deviation of 2 per cent (to be published). The analysis of blood for alcohol uses the internal-standard technique: blood, diluted with aqueous propanol, is injected directly on to the column. Interestingly, the internal-standard technique is being very widely used in forensic toxicology—examples include the use of *p*-dimethylaminobenzaldehyde for the detection of meprobamate and glutethimide (Finkle, 1967), chlorobutanol in the assay of chloral hydrate and trichloroethanol (Jain *et al*, 1967), *N,N*-dimethylaniline for amphetamines (Beckett, Tucker and Moffat, 1967), squalene for opiates and dibutylphthalate for barbiturates (Parker, Wright and Hine, 1967).

The direct injection of blood, without preliminary extraction and concentration of any poison it may contain, on to the column is another encouraging possibility for toxicologists. The injection of, say, 10 μl. of blood on to a column, which could then be screened for all common poisons by gas chromatography, is an attractive theoretical proposition. Toxic blood levels for many of the common tranquillizers are of the order of 100 μg./100 ml.; this means that about 10 ng. of drug is available for injection into a chromatograph; a detectable value. So far the technical difficulties of practical realization of this procedure have been too great, although direct injection of urine into the chromatograph for barbiturate screening is practical.

When the Home Office Central Research Establishment started work in 1967, specific research projects in areas of forensic toxicology were defined. One of these projects involved the then outstanding problem of trapping microgramme quantities of gas chromatographic effluents and preparing infra-red (i.r.) spectra from them. I stress here the microgramme quantity: forensic toxicology is one area where one cannot get a larger sample: on many occasions there just is no more body left. We have successfully solved this problem using equipment readily available in the regional laboratories and we now also routinely use gas chromatography coupled with mass spectrometry, both as a research tool and also for dealing with problems occurring in our laboratories. We use an AEI mass spectrometer, the MS 902. In just over two years, therefore, the sensitivity of absolute identification available as a routine service in the Forensic Science Service in the UK has been increased by a factor of about 10 000—from 100 g. to 1 μg., using trapped gas chromatographic fractions and micro i.r. spectrometry, and to less than 10 ng. using the gas chromatography–mass spectrometry combination. This can be called a "recent development"; it certainly stresses the value of the British system of forensic science which is nationally organized by the Home Office and so allows capital expenditure on a scale not possible for the organizational systems of many other countries. Absolute identification is vital in forensic toxicology; the scientist cannot go into the witness box and say that he "thinks" there is strychnine in the dead body. Proof is essential and it is for this reason that the Central Research Establishment has concentrated its efforts into a study of gas chromatography combined with spectroscopic techniques.

Gas chromatography–thin layer chromatography and thin layer chromatography–gas chromatography are also being used; Janák (1964) first pointed out the advantages of the combined approach; there is separation according to the number of carbon atoms by gas chromatography and according to the functional group by thin layer chromatography. Janák's ideas have been developed as a working tool in forensic toxicology by Parker, Wright and Hine (1967).

It cannot be stressed too strongly that the forensic toxicologist does not know the nature of the compound for which he is searching in his analyses. He has to preserve a healthy distrust of circumstantial evidence. The tablets scattered by the side of the bed of a suspected suicide are the ones he has *not* taken, and the child who has "accidentally" taken his mother's tranquillizers may have been helped to his death by the simultaneous murderous administration of a dozen aspirin tablets. The design of toxicological analyses is, therefore, more important if a murder by poison is to be revealed. The forensic toxicologist has always needed an analytical tool which would detect (not destroy) with high sensitivity any unusual compound in his extract and also give some lead to its identity. Gas chromatography is just a tool, but it can be used for all classes of compounds, it has very high sensitivity, retention times give clues to identity, the detectors can be made non-specific, or specific to any element at will and, by using stream splitters, the bulk of the sample can be recovered. The gas chromatography–mass spectrometry combination is probably the ideal tool for the forensic toxicologist.†

REFERENCES

BECKETT, A. H., TUCKER, G. T., and MOFFAT, A. C. (1967). *J. Pharm. Pharmac.*, **19,** 273–294.

CURRY, A. S. (1955). *J. Pharm. Pharmac.*, **7,** 969–989.

FINKLE, B. S. (1967). *J. forens. Sci.*, **12,** 509–528.

JAIN, N. C., KAPLAN, H. L., FORNEY, R. B., and HUGHES, F. W. (1967). *J. forens. Sci.*, **12,** 497–508.

JANÁK, J. (1964). *J. Chromat.*, **15,** 15.

PARKER, K. D., WRIGHT, J. A., and HINE, C. H. (1967). *J. forens. Sci. Soc.*, **7,** 162–170.

† Discussion of this paper starts on p. 145

THE USE OF GAS LIQUID CHROMATOGRAPHY IN AIRCRAFT ACCIDENT TOXICOLOGY

D. J. BLACKMORE

RAF Institute of Pathology and Tropical Medicine, Home Office Central Research Establishment, Aldermaston, Berkshire

THE investigation of an aircraft accident involving fatalities is a complex study requiring scientists from many disciplines. In addition to direct assessment of mechanical failure by engineers, considerable assistance may be given to the investigating team by pathological study of human remains. Such a study frequently throws light on the sequence of events occurring before the crash and the medical condition of the aircrew involved. Mason (1962) has extensively reviewed this aspect and no investigating team would be considered complete without the presence of a pathologist specialized in aircraft accident pathology. More recently, following an intensive study of civil air disasters, it was shown by Stevens (1968) that toxicological screening of the aircrew and selected passengers was of value in those crashes in which inconsistant or unusual behaviour of the crew before the disaster was noted. This idea has now been extended to cover complete toxicological screening of personnel from all air crashes in which pathological assistance is requested by the Board of Trade investigating team. Table I indicates the scope of such an investigation.

The problems are numerous, and considerable effort has been placed in attempting to determine therapeutic quantities of substances in *post-mortem* tissue that is frequently decomposed and contaminated. Classical spectrophotometric methods on tissue extracts are largely invalidated due to both contamination and inadequate quantity. The use of gas chromatography was therefore investigated to utilize the specificity and sensitivity of this technique.

The use of gas chromatography when searching for unknown substances against an inconsistent background reveals the severe limitation of the technique: lack of identification. This may be overcome by reducing background to negligible proportions and by identification of the column eluate.

The former is relatively easy if the concentration of the unknown greatly exceeds that of the background, but if this is not so then the latter becomes essential. Trapping of gas chromatographic eluates has been extensively studied by many investigators and recently Curry and co-workers (1968) have successfully trapped and obtained infra-red (i.r.) spectra from an initial injection of 1 μg. of barbituric acids. Such elegance and sensitivity,

TABLE I

APPLICATION OF TOXICOLOGY TO AIRCRAFT ACCIDENT RECONSTRUCTION

- Aircraft Accident Reconstruction
  - Pilot
    - Environmental toxicology — Aviation fuel; Carbon monoxide; Hydraulic fluids; Fire extinguishants
    - Deliberate medication — Nicotine; Alcohol; Decongestants; Stimulants; Hypnotics; Antipyretics, etc.
  - Sequence — in flight fire — Crew and passengers' carbon monoxide

however, has not yet been achieved for more volatile basic drugs and, to date, 10 μg. has proved the limit of sensitivity for trapping and re-injection of nicotine and amphetamine into the gas chromatograph. The use of trapping on to thin layer plates for subsequent development and identification by the standard techniques of thin layer chromatography has been studied by Parker, Wright and Hine (1967). In our experience satisfactory trapping and further identification of volatile basic drugs, even on to acid-citrate-impregnated cellulose, thin layer plates requires 10 μg. or more of injected base. The use of a coupled mass spectrometer with the gas chromatograph appears to offer the best possibility for successful identification. Mass spectra of volatile hydrocarbons can be easily obtained from 10 ng. of injected hydrocarbon but with more polar compounds the difficulty of adherence of these molecules to the glass separator situated between the gas chromatograph and the mass spectrometer has yet to be overcome. Initial studies with the use of split effluent from support-coated open tubular columns appears promising and may prove to be the method of choice.

It is in the context, therefore, of the use of the gas chromatograph for separation of an unknown component from tissue that this paper is presented.

VOLATILE COMPOUNDS

In aviation toxicology, the volatile compounds consist mainly of fuels, hydraulic fluids, antifreezing agents, refrigerant compounds and fire extinguishing materials that may be present in the environment of an aircraft cockpit and inhaled by the aircrew. Following inhalation they may be detected in blood, liver and lung depending on the time of exposure and rate of absorption from the lung into the vascular compartment.

Gas chromatographic separation of the compounds listed in Table II presents few problems. The convenient packings, Porapak Q (Hollis, 1966) at a temperature of 100–180° C, polyethylene glycol 400 at 60–100° C and

TABLE II

VOLATILE COMPONENTS LIKELY TO BE MET WITH IN AVIATION TOXICOLOGY

	Main components
Fuel (petrol) (diesel)	Paraffins (aliphatic and cyclic) Aromatic hydrocarbons Olefins Aromatic amines
Fire extinguishing materials	Halogenated hydrocarbons
Hydraulic fluids	Alcohols, ketones
Refrigerant compounds	Halogenated hydrocarbons
Antifreezing agents	Monohydric and polyhydric alcohols

Apiezon-L at 70–110° C have proved entirely satisfactory and are of value in both alcohol and drug analyses. The flame ionization detector gives adequate sensitivity following injection of the head space from 5 g. of heated blood or tissue. Electron capture detectors are of value for the halogenated hydrocarbons but have the disadvantage of being too sensitive. Those unfamiliar with their use should take precautions to avoid the presence of bottles of volatile halogenated hydrocarbons within fifty yards, even in well ventilated laboratories. Care must also be taken in removing tissue and blood from a refrigerator in a laboratory in which solvents are being used. False positives may occur due to condensation of solvents in the atmosphere on to the cold specimens.

NON-VOLATILE MEDICAMENTS

This field is vast and time does not permit a complete survey of gas chromatographic techniques for all drugs. To date efforts have been concentrated on the determination of stimulants, barbiturates and decongestants. One common factor apparent when manipulating drugs is that conditions that are suitable for 1 μg. or more of a compound are not necessarily valid for less. A good example of this is the gas chromatography of barbiturates where a column coated with silicone gum will give excellent resolution and reproducibility for 5 μg. of compound. If less than 1 μg. of barbiturate is to be identified then the behaviour of the silicone gum is unpredictable. In my experience the high efficiency polyesters, although they have disadvantages in long retention times, give symmetrical peaks and consistent retention data independent of the quantity of injected material. Neopentyl glycol adipate and cyclohexane dimethanol succinate are two such polyesters that have proved invaluable.

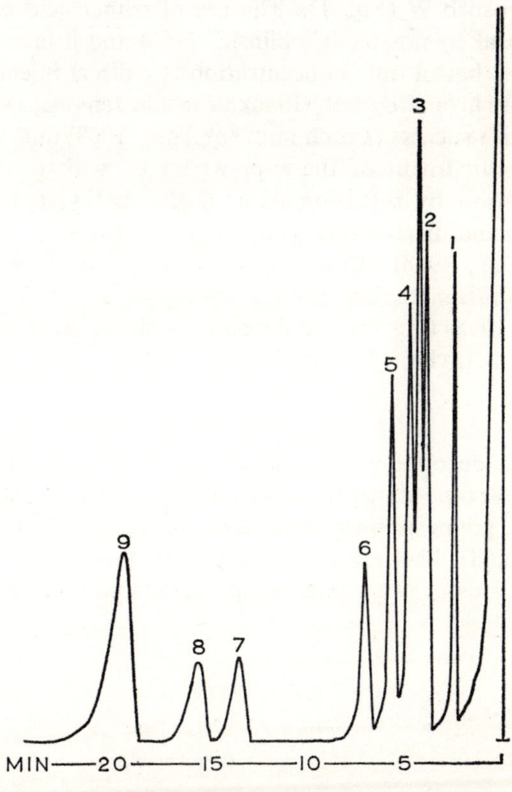

FIG. 1. Gas chromatogram of barbiturates on 0·75% trimer acid and 3% neopentyl glycol adipate (test compound acid-washed and treated with dichloromethyl silane Chromosorb W at 200° C).

Test compounds: 1. Barbitone; 2. butobarbitone (Soneryl); 3. amylbarbitone (Amytal); 4. pentobarbitone (Nembutal); 5. quinalbarbitone (Seconal); 6. hexobarbitone (Hexethal); 7. cyclobarbitone (Phanodorm); 8. hydroxyaminobarbitone (Hydroxyamytal); and 9. phenobarbitone.

Many drugs at low concentrations will adhere readily to glass and support-media, therefore all glassware and packings used in columns must be well silanized before use.

Barbiturates

Barbiturates separate well at temperatures of 200° C on 0·75 per cent trimer acid and 3 per cent neopentyl glycol adipate or 0·75 per cent trimer acid on 3 per cent cyclohexane dimethanol succinate on acid-washed and silanized Chromosorb W (Fig. 1). The use of trimer acid greatly improves sensitivity and peak symmetry (Cieplinski, 1963) and it has been possible to detect therapeutic barbiturate concentrations by direct injection of 10 μl. of urine into the gas chromatograph (Blackmore and Jenkins, 1968). Apiezon-L has been used with success (Leach and Toseland, 1968) and has been further improved by pretreatment of the support media with glycerol tristearate. Further information by retention data of the methylated derivatives will usually identify the barbiturate (Martin and Driscoll, 1966; Stevenson, 1966; Parker *et al.*, 1968). Care should be taken in preparing methylated derivatives with tetramethylammonium hydroxide as the degree of methylation of the derivatives thus formed depends on the relative concentrations of barbiturate and methylating agent.

Decongestants

Most common decongestants separate well at temperatures of 200° C on 1 per cent cyclohexane dimethanol succinate on acid-washed and silanized Chromosorb W, giving reproducible retention data and symmetrical peaks (see Fig. 2). Table III lists a few compounds and their relative retention times and clearly indicates the very wide range of compounds of high molecular

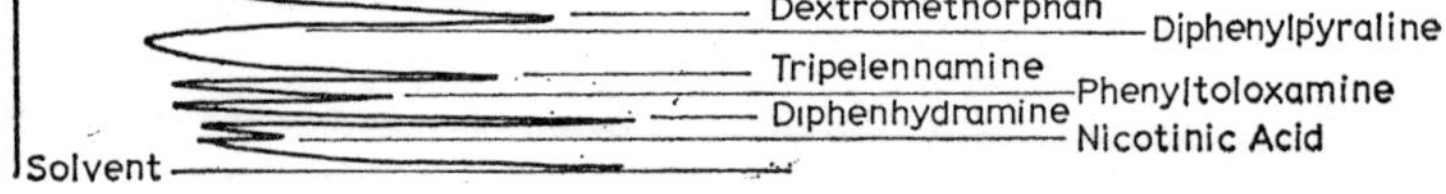

Fig. 2. Gas chromatogram of some decongestants on 1% cyclohexane dimethanol succinate on Chromosorb W (60:80 mesh) at 200° C.

TABLE III

RETENTION TIMES, RELATIVE TO NICOTINIC ACIDS, OF SOME COMMON DECONGESTANTS SEPARATED ON 1 PER CENT CYCLOHEXANE DIMETHANOL SUCCINATE AT 200°C

Compound	*Structure*	*Retention time*
Nicotinic acid	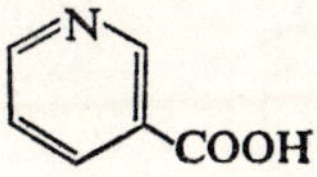	1·0
Diphenhydramine	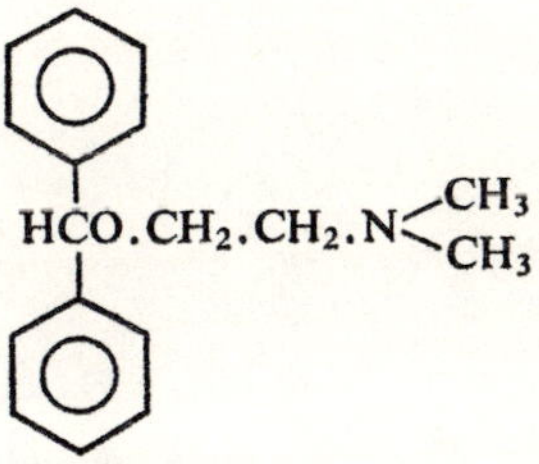	1·7
Phenyltoloxamine	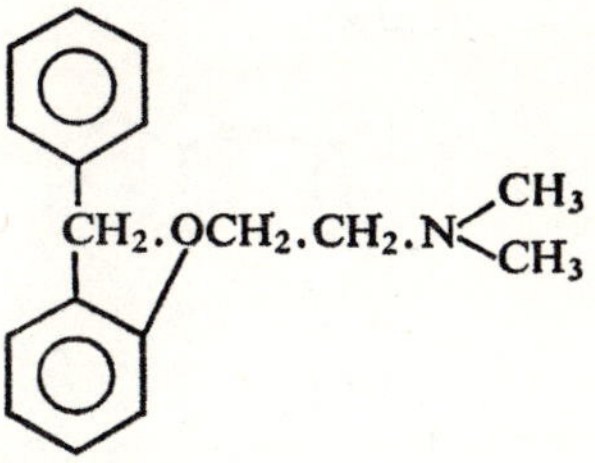	2·5
Tripelennamine	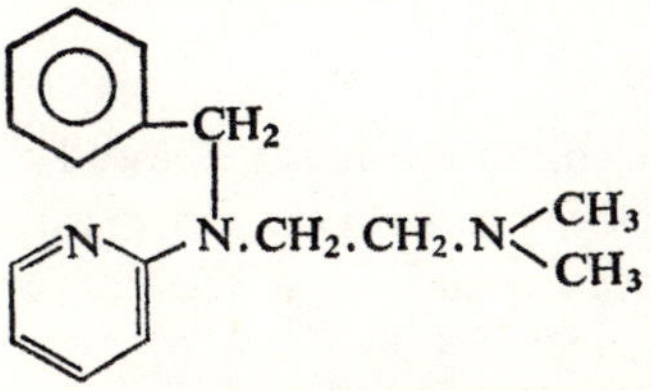	3·0
Diphenylpyraline	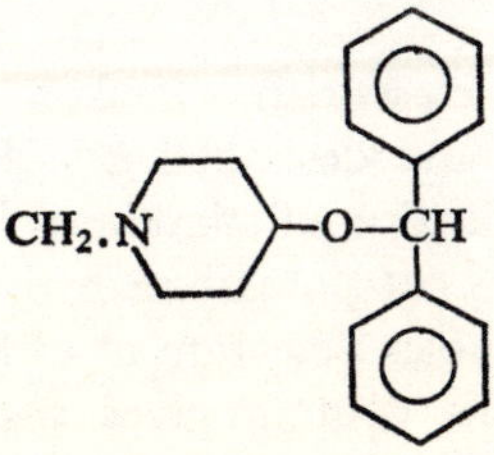	4·75

TABLE III—*continued*

Compound	*Structure*	*Retention time*
Phenindamine	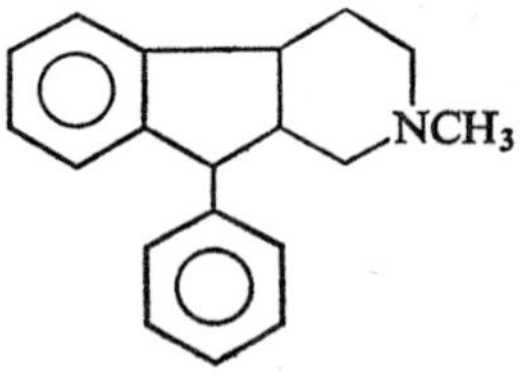	6·75
Thonzylamine	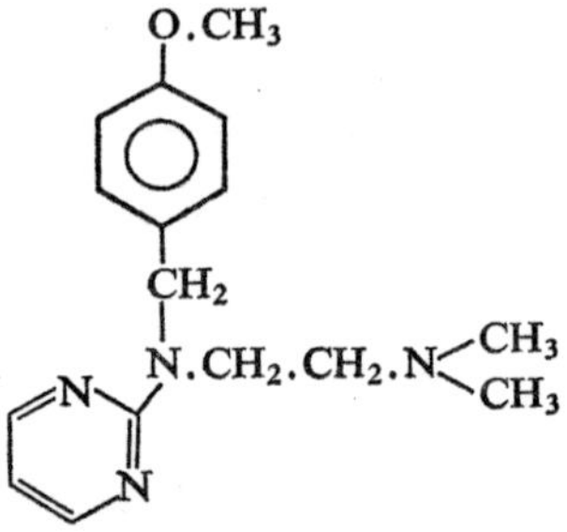	8·5
Cyproheptadine	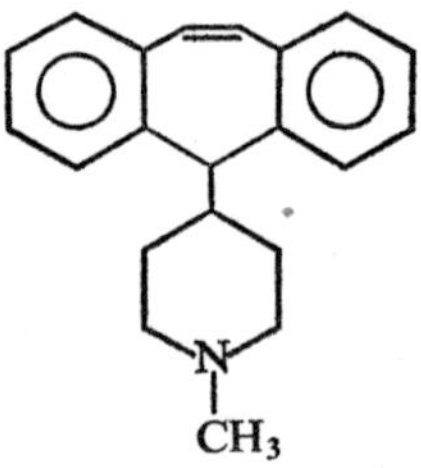	10·75

weights with which the column will deal. Should more than 10 μg. be available then mass spectrometry of the gas chromatographic eluate gives clean, well-defined mass spectra and may enable further identification of the metabolites that are present.

Stimulants

Professor Beckett and his co-workers (Beckett, Tucker and Moffat, 1967; Beckett and Moffat, 1968) have described gas chromatographic techniques for the detection and identification of stimulants in urine. The use of 10 per cent Apiezon-L and 10 per cent potassium hydroxide on acid-washed and silanized Chromosorb W enables detection of as little as 50 ng. of drug base per millilitre of urine. This column gives constant retention data and symmetrical peaks for a wide range of bases. Further identification by retention data of the acetone or N-acetyl derivative for primary and secondary amines give further evidence of identification. The application of these techniques to blood and tissue extracts has proved wholly successful and this is the method of choice for stimulants. Comparison of the chromatograph from this basic column and the antihistamine column will further identify many bases.

General

It is recommended for drug screening that at least two columns be used: one acid, for analysis of acidic extracts, and one basic, for analysis of basic extracts. In addition to selectivity, the use of one column for one type of compound prevents the formation of involatile salts at the site of injection; these may be formed should non-volatile acids or bases be present from a previous sample. A third neutral column may prove of value for rapid screening of both neutral drugs and bases that are destroyed by the potassium hydroxide or do not emerge from the basic column. Flame ionization detection is sufficiently sensitive for most purposes.

Ethyl alcohol

Gas chromatographic techniques for the determination of ethyl alcohol are well documented and the method of Curry, Walker and Simpson (1966) has proved of great value.

It is impossible to overemphasize the importance of a gas chromatographic technique for the determination of alcohol in decomposing blood and tissue.

TABLE IV

VOLATILE ORGANIC COMPOUNDS DETECTED IN DECOMPOSING TISSUE

Alcohols	Methyl, ethyl, isopropyl, propyl
Aldehydes	Acetaldehyde, propionic aldehyde, isobutyl aldehyde
Ketones	Dimethyl, methylethyl, diethyl
Acids	Acetic, propionic, n-butyric, iso-butyric

Bacterial contamination may cause falsely elevated values, and Table IV lists compounds that can interfere with analysis and that have been isolated and identified from non-sterile tissue.

It is essential therefore to use two columns for alcohol determination; polyethylene glycol 400 and Paropak Q are recommended (Blackmore, 1968). Urine is the body fluid of choice for such estimations, since the occurrence of false results that may occur due to bacterial contamination is least when samples are taken from urine.

Carbon monoxide

The use of molecular sieve 5A allows carbon monoxide to be separated from other gases. Elutriation of the molecular sieve with water (Farre-Ruis and Guiochon, 1963) considerably enhances the performance. Treatment of blood or tissue extract containing blood with alkaline ferricyanide solution liberates the carbon monoxide from the haemoglobin and the gas can then be quantitatively transferred via a gas loop to a gas chromatograph. Electrical integration of the katharometer response and comparison with a standard enables an accurate determination of the quantity of carbon monoxide per aliquot of fluid. From the ratio of this quantity to the total haemoglobin concentration a percentage saturation may be calculated. This technique has proved invaluable in this work, the results having considerably assisted accident investigation teams to reach a conclusion.

CONCLUSIONS

The use of gas chromatography has made the screening of aircrew fatalities for therapeutic medication, alcohol, carbon monoxide and environmental contamination a practical possibility. An investigation of this nature, though largely negative, assists the aviation pathologist to assess the *ante-mortem* condition of aircrew, and has provided direct evidence of accident causation in at least one air disaster in the past year.

Acknowledgements

My thanks are due to Dr. A. S. Curry, Director of the Home Office Central Research Establishment, for his constant advice and encouragement during this investigation, and to Mr. R. Jenkins for invaluable technical assistance.

I wish to thank the Royal Air Force Director General of Medical Services for permission to present this paper.

REFERENCES

Beckett, A. H., and Moffat, A. C. (1968). *J. Pharm. Pharmac.*, **20,** Suppl., 48s–50s.

Beckett, A. H., Tucker, G. T., and Moffat, A. C. (1967). *J. Pharm. Pharmac.*, **19,** 273–294.

Blackmore, D. J. (1968). *J. forens. Sci. Soc.*, **8,** 2–3, 73–78.

Blackmore, D. J., and Jenkins, R. (1968). *J. forens. Sci. Soc.*, **8,** 1.

Cieplinski, E. W. (1963). *Analyt. Chem.*, **35,** 256.

Curry, A. S., Read, J. F., Brown, C., and Jenkins, R. (1968). *J. Chromat.*, **38,** 200–208.

Curry, A. S., Walker, G. W., and Simpson, G. B. (1966). *Analyst, Lond.*, **91,** 742–743, 1088.

Farre-Ruis, F., and Guiochon, G. (1963). *J. Gas Chromat.*, **1,** 33.

Hollis, O. L. (1966). *Analyt. Chem.*, **38,** 309.

LEACH, H., and TOSELAND, P. A. (1968). *Clinica chim. Acta*, **20**, 195.
MARTIN, H. F., and DRISCOLL, J. L. (1966). *Analyt. Chem.*, **38**, 345.
MASON, J. K. F. (1962). *Aviation Accident Pathology*. London: Butterworth.
PARKER, K. D., WRIGHT, J. A., HALPERN, A. F., and HINE, C. H. (1968). *J. forens. Sci. Soc.*, **8**, 2–3, 125–140.
PARKER, K. D., WRIGHT, J. A., and HINE, C. H. (1967). *J. forens. Sci. Soc.*, **7**, 162–170.
STEVENS, P. J. (1968). Ph.D. Thesis, University of Birmingham.
STEVENSON, H. W. (1966). *Analyt. Chem.*, **38**, 1948–1949.

DISCUSSION

Lipsky: I congratulate you both on bringing the latest technology into forensic medicine. This confirms how suitable for forensic identifications these new techniques are. One of the most exciting areas of progress in this field is our increasing ability to make smaller, less expensive and more versatile mass spectrometers. A quadrupole mass spectrometer has a growing market in the US and will shortly be available in the UK at a cost of less than £8000. This machine can couple the open-support tubular columns directly—that is, without a molecular separator—to the mass spectrometer. There have also been intriguing developments in the columns themselves; any type of stationary phase can be introduced now, including molecular sieves on the column. These can tolerate sample sizes of up to 500 μg. The quadrupole mass spectrometer is compact and so sensitive that it detects amounts as small as 0·1 ng. in the ion source. By coupling to an open-support tubular column this machine can analyse the entire effluent without any sample splitting. It tolerates helium flows of the order of 2–3 ml./min. with negligible loss of sensitivity. Other ways of using this excellent device might include ancillary techniques utilizing pyrolysis and molecular separators, followed by identification by mass spectrometry. This exceedingly powerful instrument would be extremely useful to forensic toxicologists.

Blackmore: We have tried coupling a gas chromatograph with the AEI mass spectrometer (model MS 9) but find that the actual geometry of the apparatus precludes any chance of an accurate result; it is impossible to fit the chromatograph on to the end of the MS 9 because of difficulties connected with alignment.

Purnell: The MS 10 mass spectrometer might suit you better because one can build one's own inlet system on to it, with consequent higher sensitivity in the chamber. The MS 10 is about one tenth the price of the MS 12.

A. Curry: Instrumentation is still only one part of the problem. Even when we have suitable techniques and outputs we still need a huge collection of control spectra with which to match the components of the unknown sample. We usually have little idea of the nature of the compound under test although we do know the retention time. In i.r. analyses, for example, a working collection of the spectra of two to three thousand compounds is essential and has now been produced at our Establishment (Curry, Read and Brown, 1969). We need this control collection in order to retrieve the unknown compound rapidly. Thus one of the first stages in this programme must be to build up such a mass-spectra library.

Brooks: What range of compounds is covered by your systematic analyses? The compounds you and Mr. Blackmore mentioned were all stable in themselves to gas chromatography, but one cannot rely on a poisoner's always using volatile chemicals! Have provisions been made in your present system for studying compounds that have to be made into derivatives because they are either non-volatile or volatile but unstable?

A. Curry: Any system of poison detection must devise ways not only of finding one particular poison out of the thousands that might be present but also of investigating the normal metabolism of the body to detect possible abnormalities. A biochemical lesion might give a lead about the type of poison that had been used. In this very wide field, full systematic toxicological analysis must consider cholinesterase inhibitors together with arsenic, for example, and yellow phosphorus with heroin. (See Curry, 1969).

Lowe: Our laboratory may be asked to determine anaesthetic concentrations at post-mortem. If the post-mortem is at our own hospital we can complete our tissue analyses before the autopsy is finished. But when we need to retain samples (for example if tissue is mailed from other hospitals) it is convenient to use flux-free solder tubing, particularly for transporting volatile materials.

Hill: We have used plastic disposable syringes for collecting our blood samples, mainly because they are so cheap. Unfortunately Dr. Denison Davies in our laboratory (unpublished material) found that significant amounts of nitrous oxide are lost from these syringes, which precludes our using them at the moment. I used old-fashioned tuberculin syringes (glass with metal plungers) in my original work on blood gases, so I had not met this difficulty until recently. Have you met this problem with carbon monoxide, Mr. Blackmore?

Blackmore: We mix the blood sample with alkaline ferricyanide for about 15 minutes and then inject into the chromatograph. I have repeated these analyses 12 hours later, with the same aliquots of blood and ferricyanide, and obtained an exactly similar chromatogram. I have never found any significant loss of carbon monoxide but I may just have been lucky or had a particularly good batch of plastic syringes. And I have never tried to conserve blood for the subsequent detection of nitrous oxide.

Lowe: We studied the stability of volatile anaesthetics in blood, using polyethylene disposable syringes, and found a loss of 30–40 per cent in 24 hours. In contrast, halothane concentrations in rat brain were stable for 6 months when preserved *in situ* at a temperature of 4° C.

Blackmore: One important danger when samples are stored at a temperature of – 20° C and then taken out to thaw in the laboratory is that solvent readily condenses on the specimen. It can be extremely misleading when one is looking for very small amounts of volatile compounds to find that large volumes of similar compounds have condensed on the surface of the tissue when it was thawing out.

Lowe: Distilled water in bottles will dissolve anaesthetic vapours from the surrounding atmosphere. We frequently have to boil our water before making a standard.

Payne: Interestingly, awareness of these phenomena is due to the accuracy and reproducibility of gas chromatographic techniques. Many of these problems were not even considered until we started to use gas chromatography.

Lowe: One must also guard against contamination when anaesthetics are being analysed in the clinical situation; when ether or isopropyl alcohol are used to prepare the skin, significant quantities can be absorbed and produce fairly high peaks on the chromatogram.

A. Curry: Professor Lipsky, you mentioned pyrolysis techniques. It is routine procedure in forensic science to use pyrolysis for the comparison of paints and plastics. Another useful technique involves the use of one gas chromatograph to purify and separate our compounds, with a pyrolyser introduced into the circuit and preceding another chromatograph. Do you have any experience with this sort of technique?

Lipsky: A gas chromatograph and a pyrolysis unit have been arranged in series and used by Simmonds, Shulman and Stembridge (1969). The pyrolysis unit is a short compact oven, modelled on the original design of Keulemans (Keulemans and Cramers, 1964) who used a gold capillary tube in their apparatus. This system gives a performance somewhere between that of the gas chromatograph and of the mass spectrometer. When a sufficient library of pyrolysis patterns for a wide range of substances has been built up, this technique could be a good compromise between the gas chromatograph and the mass spectrometer.

A. Curry: This would be very useful to us. Accurate and reproducible results are difficult to obtain from pyrolysis in a gas-flow stream.

The use of miniaturization in columns, detectors and amplifiers would be another valuable development for forensic toxicologists working in the field. In the investigation of aircraft accidents, for example, one may want to detect minute amounts of toxic materials in an aircraft cockpit but one needs a two-ton truck to transport the chromatograph. How soon will the manufacturers be able to supply us with a Sparklet-bulb-type gas-cylinder device and a gas chromatograph that can be held in the hand?

Scott: What materials are you looking for in the aircraft?

Blackmore: Carbon monoxide is still one of the most important compounds we meet in aircraft accident toxicology. Attempts to detect carbon monoxide at the site of the accident may involve a search over several miles of countryside and the examination of small, scattered pieces of human tissue. A method giving a reliable quantitative estimate of the carbon monoxide content of the tissue at the site of the accident would be invaluable. Another problem in this work concerns the detection of small quantities of volatile materials (such as aviation fuels and hydraulic fluids) in aircraft cockpits. A pilot may develop visual difficulties (multiple vision for example) after taking off. After he has landed we are called in to investigate the problem in an aircraft that may be several hundred miles away from our laboratory. Ideally, we require large volumes of blood and urine, and liver and kidney biopsies, because the concentration of carbon monoxide in the tissues is so small, but obviously it is not practicable to obtain such specimens. A pocket kit that could be used on the air in the cockpit and on the pilot while the toxic materials were still present would be extremely useful.

Scott: The analysis of aviation fluid and similar materials in the air is relatively simple; satisfactory miniaturization of the appropriate apparatus into a portable package merely awaits the money needed for its development.

Lowe: W. F. Whilhite of the Jet Propulsion Laboratory, Pasadena, has described a thermal detector measuring only 100 × 100 × 150 mm. This consists of a sampling valve, columns, detectors, a carrier-gas container, pressure regulators and electronic apparatus for analogue to digital readout. The unit is battery operated and weighs 100 grammes (National Aeronautical and Space Administration [NASA] technical brief 66–10182, Springfield, Virginia, 22151).

Purnell: The chromatograph designed by NASA for the first soft landing on Mars weighs only about 2 kg. Is this the sort of equipment you need, Mr. Blackmore?

Blackmore: Yes. This is exactly what we want for our carbon monoxide estimations.

Lipsky: The problem of sampling is crucial. Many thousands of litres of air must be moved from an aircraft cockpit through a small trap to concentrate the minute amounts of carbon monoxide that may be present; this puts quite a strain on the detecting capabilities of the instrument and some sort of concentrating device is essential. After concentration, the trapped gases can be blown into a chromatograph and all other components more or less excluded.

Blackmore: I agree that the detection of very small concentrations of toxic materials from large volumes of air is a problem. But a far greater difficulty, as I have said, is the analysis of small pieces of tissue that have to be transported many miles to the laboratory. And if the initial analysis does not produce the required information, we have to travel back to the site of the accident for another experiment.

Concerning carbon monoxide, I would appreciate any ideas on how to record data from a miniature chromatograph. I have seen microkatharometers, and in our laboratory we have made a miniature molecular-sieve column consisting of 150 mm. of thin copper tubing (diameter = 3 mm.) packed with molecular sieve 5A. This efficienctly separates carbon monoxide from oxygen and nitrogen. We could certainly miniaturize an amplifier for this, but how would we record the output?

Purnell, Lipsky and Scott: On a tape recorder.

Purnell: To monitor low levels of hydrocarbons a "total hydrocarbon analyser" (that is, just a flame detector) is quite adequate. Since most hydrocarbons give a nearly uniform response per g.atom of carbon the results can be expressed in parts per million of equivalent carbon. Such a detector system need be only about the size of a cigarette packet.

Scott: There is no scientific reason why the chromatograph should be so large. In fact a great deal of space is wasted around components that have already been miniaturized at small cost. The fantastic extremes of miniaturization resulting from the work of NASA in the US have been expensive of course, but reasonable miniaturization, to the same order of size as the modern transistor radio, is perfectly possible for chromatographs. Such equipment would easily fit into an aircraft cockpit.

Lipsky: I would like to re-emphasize the usefulness of the quadrupole mass spectrometer for the forensic toxicologist. Another advantage of this instrument is that screening can be done very quickly if some of the volatile material is allowed to enter directly into the machine, thus producing a range of molecular weights. If one finds a molecular range around a molecular weight of 600, one knows at least that the compounds under test may not be sufficiently volatile to be analysed with gas chromatographic techniques.

The sensitivity of the quadrupole mass spectrometer is in the same range as that of the flame ionization detector although the linear dynamic range of the former instrument is much less than that of the latter.

Scott: Do you have a problem about the sensitivity at which you can detect carbon monoxide, Mr. Blackmore?

Blackmore: Rarely, but it depends on the amount of carboxyhaemoglobin available for the analyses.

Scott: Is there usually a reasonable amount of tissue for you to examine?

Blackmore: This depends entirely on the circumstances of the accident. If one has only a few grammes of burnt muscle to work with insensitivity can be a major problem.

Scott: This problem might be resolved in the way I described in which carbon dioxide or carbon monoxide can be converted to methane using Cropper's technique (Cropper, Heinekey and Westwell, 1967) and a flame ionization detector.

Blackmore: The main problem about sensitivity is not to decide how much carbon monoxide is present but what it is related to. The concentration of carbon monoxide in blood or tissue fluid may be 10 μl./ml. blood, but this volume comes from an unknown amount of haemoglobin. We deal with this difficulty by determining the ion content with atomic absorption spectrophotometry, thus calculating the haemoglobin, and then confirming by standard measurement of cyanmethaemoglobin after dilution with a cyanide solution; but at very low concentrations even these methods are unsatisfactory.

A. Curry: A rapid and sensitive technique for the identification of sex hormones in blood stains would be another useful tool for the forensic scientist. Suspects in a murder inquiry may have blood stains on their clothing. If we could find out if the blood comes from a man or a woman we would be able to eliminate several suspects at an early stage.

Sjövall: We have studied certain steroid sulphates in blood with gas chromatographic-mass spectrometric methods.† Gas chromatography can be combined with electron capture detection (see Eik-Nes and Horning, 1968) but electron capture detectors are generally said to be difficult to use in routine steroid analysis because they become contaminated by the samples. We are now starting some work using the mass spectrometer as a detector. Detection with this apparatus can be very specific and there is no theoretical reason why it should not be possible to detect picogramme quantities.

Payne: Are the problems you mention technical?

Sjövall: The adsorption and decomposition of compounds on the column—

† See also pp. 203–205.

column technology—is one difficulty. Another is the isolation of material from the biological sample. It is relatively easy to inject the sample into the combination instrument and focus on a particular ion that is as specific as possible for the test compound; the problem is to reach the stage at which the compound can be injected on to the column.

Payne: In this work, in contrast to forensic toxicology, at least we know what we are looking for.

Sjövall: This is obligatory. One tries to make a derivative of the steroid to be determined which gives an ion with high intensity and of relatively high mass. The current of ions of this mass is followed, and the specificity of the method depends on the mass of the ion chosen and the retention time of the gas chromatographic peak that it produces.

A. Curry: In my experience, when the technological problems of identifying concentrations of the order of 10 μg./100 ml. are solved, one has to start looking for a different technology for detecting amounts of 1 μg. per cent. At nanogramme and picogramme levels we meet difficulties due to artefacts caused by the tissues becoming contaminated with volatile materials picked up in the laboratory. Similar contamination occurs when female workers determine sex hormones. At certain stages of the menstrual cycle the analyses of such minute amounts of material can be upset by the girls just picking up the containing flasks.

Lipsky: Similarly, in the analysis of meteorites, Dr. Paul Hamilton of the Dupont Research Laboratories (personal communication) has demonstrated that sub-microgramme amounts of amino acids can be detected on the surface of a meteorite after it has been touched once by one finger.

Sjövall: We have also met this problem, which is one of the reasons why the detectors must be made more specific if they are to be more sensitive. The chief contaminant in our laboratory is dioctyl phthalate.

S. Curry: Concerning the apparent difficulty in using the electron capture detector in routine quantitative analysis, I have successfully used this detector for the routine analysis of chlorpromazine (Curry, 1968). Several important problems arose and were overcome before the method (extraction, and gas chromatography of the concentrated extract) became reliable. The response of the instrument was greatly influenced by contamination by the solvents, but by suitable clean-up procedures I was able to estimate chlorpromazine reliably in plasma at concentrations as low as 10 ng./ml. Chlorpromazine, of course, is a relatively non-volatile drug, and the estimations had to be done at a high temperature (250–300° C). This gave rise to loss of stationary phase (bleeding) of material from the column packing, and column bleed (like chemical impurity) adversely affected the detector. This problem was overcome by the use of OV 17 (fluorosilicone polymer, Applied Science Laboratories), from which bleed is minimal, as column packing.

Voltage control was a major problem at first but was eventually overcome quite simply. Graphs of detector sensitivity against voltage (^{63}Ni radioactive foil; continuous d.c. in concentric tube detector) demonstrated wide variation in response. The response of the detector was not linear at the lower voltages but at high voltages it was. Variation of the response to

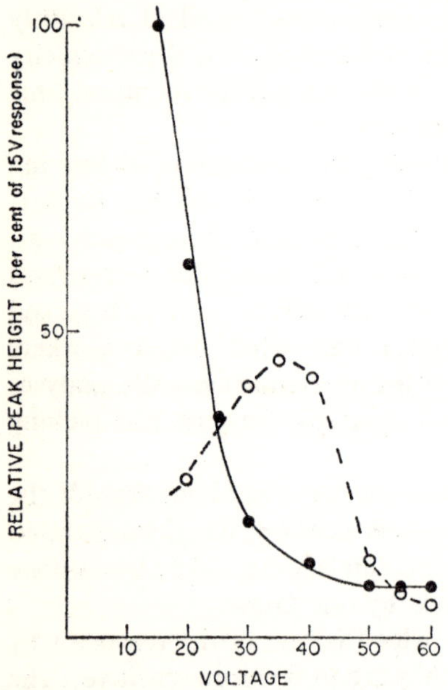

Fig. 1 (*S. Curry*). Relative response of the nickel-63 detector to a 20-ng sample of chlorpromazine at various voltages. Nitrogen (●——●) or argon (○---○) as the carrier gas.

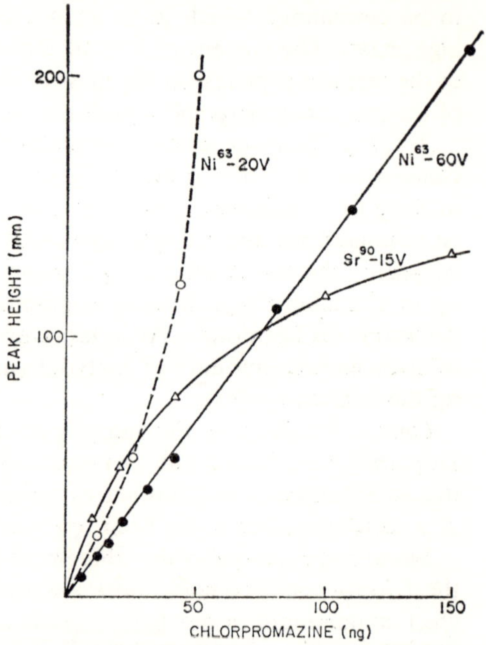

Fig. 2 (*S. Curry*). Responses of various electron capture detection systems to varied quantities of chlorpromazine.

(Curry, 1968; reproduced by kind permission of the editors, *Analyt. Chem.*)

repeated samples of the same size was observed in the areas of the response-voltage curves in which sensitivity changed markedly with small changes in voltage. "... This variability may have resulted from insufficient voltage control, the significance of which would have been greatest in the areas of the voltage response curve showing marked variability of response with small changes in voltage. The use of 60 V did not reduce the overall sensitivity of the method for this was governed by the relative magnitude of the experimental signal and the signals of solvent and instrumental noise. These three signals were affected equally by the changes in voltage, so the more stable, high voltage conditions could be used, and, with a higher electrometer gain setting than is suitable at lower voltages, the overall sensitivity was unchanged". (Curry, 1968.) (See also Figs. 1 and 2, from Curry, 1968.)

Lipsky: Did you clean the anode in the source?

S. Curry: Many times in the early stages. But finally, by taking the precautions I have described, the need for this was minimized. I see no reason why the electron capture detector should not be freely used provided that the material is clean, column bleed is low and the detector voltage is adequately controlled.

Hill: We have found that control is smoother when the polarizing voltage for the electron capture detector is pulsed. Detectors are only physical devices but a residual area of black magic still shrouds "the column". Obviously contamination gives trouble but the operation of the detector itself should be straightforward. Our gas chromatograms are sigmoid in shape, with a workable plateau in the middle of the curve; pulsing linearizes the calibration curve. In general, many of our difficulties could be eliminated if we understood properly how the detectors work.

Lipsky: When one is analysing biological samples, the stationary phase in the columns, maintained at a temperature of 200° or 250° C, will invariably break down or bleed to some extent. The terms of reference may change by a factor of 100 or 1000 when dealing with polyester or one of the more thermostable silicone columns. But with a technique combining gas chromatography and mass spectrometry one realizes this for the first time because this bleeding is now significant as base signal instability (background "noise") whereas with other systems it had always been ignored. With proper conditioning and elimination of the monomers and dimers from particular types of stationary phases this bleed can be minimized.

Martin: Could one separate the sample from the material bleeding from the column by trapping and diverting to a short column with a negligible bleed?

Lipsky: This would be possible but very difficult. The properties of the materials that bleed from the column are such that they tend to keep with the sample.

Martin: The materials bleeding from the column must cover a range of properties, and a small following column, put on at the time the sample is expected, would clean up the picture.

Gray: We met this difficulty in our work on the recovery of lipids from gas chromatographs. We were recovering dimethyl acetals of long-chain aldehydes separated on a normal ethylene glycol adipate column, and were able to obtain extremely efficient cleaning by passing the effluent through a small column of silicic acid. The effluent vapours were condensed from the gas stream by cooling traps. The condensate was dissolved in benzene and washed through a small silicic-acid column which retained the bleed from the adipate but not from the dimethyl acetals. When polyester columns are used silicic acid is very effective in removing contaminating bleed from the lipids separated by the columns.

Lipsky: I predict that the trend in chromatography will soon turn towards liquid chromatography again. Liquid–solid systems in which derivative forms and samples in picomole quantities can be analysed will become commonplace. Additional techniques with thin coatings of special stationary phases that do not bleed in the liquid phase will be evolved. We shall eventually be able to analyse steroids, including the sex hormones, with these sorts of techniques.

Scott: I agree with your predictions. We are in a transitional stage between the techniques of gas chromatography and liquid chromatography, and many workers are still trying to use gas chromatographic techniques beyond the practical limits of their operation. Until two or three years ago there was no alternative to gas chromatography, but recent work on liquid chromatography (see for example, Zlatkis, 1967, 1969) has shown that this is no longer the case. The extent to which we shall continue to use gas chromatographs at high temperatures and tolerate all the associated problems is limited. Liquid chromatographs can now be made to give resolutions and analysis times similar to those of gas chromatographs. In this transition period, research on the separation of picogramme amounts of mixtures by liquid chromatography might well be more rewarding than the study of problems connected with the detection of these small quantities on gas chromatographic columns at high temperatures.

Sjövall: I do not agree that gas chromatography is becoming less useful. It is not a difficult step in steroid analysis to make derivatives, certainly it is much easier than the many preliminary purifications that are needed in this work. The quantitative detection of very small amounts is the limiting factor and this applies to gas and even more to liquid chromatography. With liquid-gel chromatography we get plate heights approaching 0·1 mm., which is sufficient for several steroid separations (Sjövall, Nyström and Haahit, 1968). However, for analytical separations, I still prefer gas chromatography.

Scott: But when the point is reached at which the concentrations to be detected are giving signals that are of the same order as the noise produced from the column system, sensitivity can be improved no further. In other words, when the partial vapour pressure in the column at the required temperature gives a background signal that varies to the same extent as the signal we are looking at, no more progress can be made.

Sjövall: I do not agree. If one uses ions specific for the compounds being analysed one can select and work in an area with very little background noise. The mass spectra of silicones, for example, show few ions which interfere with the fragment ions that are suitable for the detection of steroids. But if one found an ion source which gave a better ion yield one would increase sensitivity to such an extent that one would again meet problems with adsorption and decomposition on the column.

Scott: We are now suffering from a "detector-column syndrome" in which we chase detector and column performance alternately. It would surely be more fruitful to start again with liquid chromatography. It can be shown both practically and theoretically that we are near to the limits of sensitivity and operating temperatures in gas chromatography.

Lipsky: Both systems can still be useful. Future development with the liquid chromatograph might be towards making a specific derivative for specific types of detection systems—a derivative with an intense u.v. absorption moiety, for example, might be useful.

Purnell: Support for the arguments that the usefulness of gas chromatography cannot be much further extended is, first, the fact that detectors are already stretched for sensitivity and, second, that higher column temperatures are of little value because solvents become more nearly ideal as the temperature increases and hence lose their selectivity. Thus longer columns are needed at higher temperatures and longer columns mean greater sample dilution which puts further demands on the detector. The attraction of liquid–liquid and liquid–solid chromatography, operating at low temperatures, is that they still offer non-ideality of the sorption process, thus allowing better separation ratios, shorter columns and less dilution of these very small amounts before they reach the detector. Investment in the development of increased detector sensitivity may no longer pay off; after all, fifteen years of intensive research into the design of detectors has not yet produced an ideal instrument.

Sjövall: I am still not convinced that we are approaching the limits of usefulness of gas chromatography. In the steroid field, at least, considerable development is still possible.

Lipsky: This is true for steroids. But one should recognize the physical limits of these techniques and, when they are reached, look for other methods.

Scott: Especially when the technique of liquid chromatography looks so promising.

Lipsky: Silanization is another neglected field for study. What happens on the surface of a column operated at a temperature of 200–225° C? Although we maintain that the columns must be silanized to prevent active sites from adsorbing the sample and thus vitiating quantitative analysis, we have few good data about the properties of these sites at high temperatures. If a column containing the 1 per cent solution of stationary phase is silanized and maintained at 225° C, are the effects of silanization on the surface of the support material the same as these effects at room temperature or at 100° C?

Purnell: At 250° C they probably are similar. But the silyl-ethyl bond becomes relatively unstable at a temperature of about 300° C. The kinetics of such decompositions at high temperatures have not been properly studied but knowing, approximately, the silyl-oxygen bond strengths, one would expect the silanization technique to be self-defeating when columns are used above a temperature of 300° C. However, at temperatures of 250–350° C the hydroxyl groups, the oxygen, and the free radicals on solid surfaces would be losing their potency as adsorption sites. In practice, no molecule bigger than about C_{16} will have any stability at 300° C. At the high temperatures suggested by some advertisements for gas chromatographs (even up to 400° C), what but the simplest of chemicals could possibly be put through the columns?

Lipsky: In an unreported experiment in 1963, I collected the peak from injected [^{14}C]cholesterol on a silanized column. But when I measured the residual radioactivity I found that about 25 per cent of the counts were still in the column.

Purnell: We have recently tried to invent column liquids which exhibit no change of retention with temperature over short temperature ranges. Such liquids might be useful for various purposes. Amongst other compounds we thought that liquid crystals might show this behaviour. Using a column of cholesterol we found that when it had once been heated to a temperature of about 130° C it then gave temperature independent retention for many volatile substances over the range 105–125° C. In other words, temperature swings within this range did not alter the chromatogram. We were surprised that this cholesterol column worked so well, but differential thermal analysis showed that the cholesterol had been degraded to cholestadiene and water in the preliminary heating on fire-brick at 130° C.

Brooks: Those of us who work with steroids try to avoid putting them down hot columns packed with active catalysts!

Purnell: I must point out that what you call the active catalyst is, in this example, silanized fire-brick, which could be regarded as providing the lower limit of solid surface activity in gas chromatography.

Sjövall: Dehydration of free cholesterol on the columns may take place. This is much more pronounced with the 3-hydroxy-Δ^4-sterol, which loses water in the flash heater, forming cholestadienes (Okerholm, Brecher and Wotiz, 1968). Silyl ethers of allylic steroid alcohols have also been reported to decompose (Kandutsch, 1967) but we found them stable when we tested them. These differences are probably due to differences in column technology. By taking mass spectra through the entire course of the chromatograph one can often check if degradation products are appearing.

Lipsky: This check on technique is important.

Sjövall: Different columns behave differently, and catalysts do not have to be deliberately introduced. Some columns seem to be made catalytic during preparation although we do not know how this happens.

Purnell: Differential thermal analysis can bc helpful over this problem since a sample of the packing one proposes to use in the column as well as the steroids being tested may be studied, interaction and other thermal effects readily detected and the corresponding temperatures defined. The experiment only takes about twenty minutes and is fully automatic.

Lipsky: The compound with which we coated our column changed its chemical structure in the first 12–24 hours but we do not yet know how this happens.

Purnell: The differential thermal analyser—a powerful tool that produces a great deal of well-defined information—can help to solve this problem too. A good thermogravimetric analysis gives data about deactivation, decomposition, the types of degradation occurring in the sytem and so on. It is well adapted for exactly this type of study.

Warren: Concerning the stability with respect to the temperature of the deactivated substrate, we injected water vapour into columns at 130°C, deactivated with hexamethyldisilazane and found that this procedure produced additional peaks. I do not know what degradation products were appearing but water was presumably reacting with the hexamethyldisilazane producing a peak at the detector.

Mitchell: Could a technique in which i.r. analysis, using ultra-micro quantities on pellets, is used be useful for detecting fat-soluble substances such as the steroids, where good results may be obtained using attenuated total reflectance methods? Has anyone any experience in the use of these techniques for estimating sub-microgramme amounts?

Lipsky: How specific is attenuated total reflectance in detecting useful characteristics?

Mitchell: The band positions and shapes of properly recorded spectra are identical to those of transmission spectra.

Lipsky: We are using i.r. analysis less as our gas chromatographic techniques become more sophisticated. We get as much information from running our samples on several different column systems as we do from i.r. pellets. The interpretation of results from i.r. analyses of certain biological materials is becoming a problem.

Mitchell: But for two similar compounds one can still complete many chromatographic runs in different systems and all will show identical R_F values. Only the production of an i.r. spectrum will show the difference.

Lipsky: This is certainly true for the thousand or so compounds now available. We must try to use the most efficient and appropriate gas chromatographic columns, polar or non-polar, when attempting to separate structurally similar compounds.

Mitchell: We have been convinced on several occasions when we were looking for new steroids using thin layer and gas chromatography that a compound being tested was the same as a standard; yet the i.r. spectrum would show sometimes major sometimes minor differences.

S. Curry: Examples obviously exist of pairs of compounds with similar chromatographic properties and different i.r. spectra, and there are also pairs of compounds with similar i.r. spectra and different chromatographic properties. The use of both these (and additional) techniques is surely the best procedure.

Lipsky: This is why the preparation of the sample is so important. One must start with some knowledge of the class of compounds one is dealing with and then try to separate a few particular compounds from thousands of others.

Brooks: On occasions when we do not have prior knowledge, as (often) the forensic toxicologist does not, the use of derivatives should not be overlooked. Making a derivative of the parent unknown may be useful not only for our particular purpose but also because it teaches us a great deal about the nature of the materials. The fact that many compounds show similar properties in one or other test system means that one should use every available analysing technique.

Blackmore: The dangers of using only one technique are illustrated by the following example: we were recently asked to examine a urine sample for the suspected presence of amphetamine. In our normal amphetamine system a large gas chromatographic peak, corresponding exactly to methyl amphetamine, appeared (Beckett, Tucker and Moffat, 1967). We had plenty of urine so we continued our tests with thin layer chromatography and produced a ninhydrin reacting spot, again corresponding exactly to methyl amphetamine. A more specific confirmatory test for amphetamine is to observe the change in retention time of the acetyl derivative of methyl amphetamine. We failed completely to do this with our compound, which is still unidentified. Grave dangers of inaccurate identification exist if one uses only one method.

Lipsky: Dr. Curry, how do you detect insulin when a massive dose has been given subcutaneously?

A. Curry: The first successful detection of murder by insulin poisoning was in 1957 (Birkinshaw *et al.*, 1958). A woman had been drowned in the bath; there was a previous history of vomiting, dilated eye pupils and sweating and, although we found no toxicological abnormalities at post-mortem, hypodermic needle marks had been found in buttock tissue and we were convinced that some toxic substance had been injected. The problem was what to do with this tissue. We were misled at first by the fact that the blood sugar levels were higher than normal. Then we realized that we were testing mixed heart blood and, because liver glycogen had been broken down, blood sugar levels taken at post-mortem meant nothing. In this case we eventually made the correct diagnosis of the cause of death by extracting the protein from the buttock tissue and assessing it biologically in mice and in rat diaphragm preparations.

Lipsky: That must have been a laborious procedure.

A. Curry: This can be more simply done today with radioimmunological assay techniques. In this particular case, as well as looking for insulin itself, we also had to try and determine which type of insulin had been used according to the preservatives in insulin solutions (cresols, phenols, methyl-parahydroxybenzoates and so on).

Lipsky: How do you investigate death from anaphylaxis due to a foreign protein?

A. Curry: This is difficult. Death from wasp stings may never be correctly diagnosed (Richey, J. W., Kehoe, J., and Stauch, J. E. Material presented in *Proc. III Int. Med. Forens. Path. Meet.* [London, 1963], unpublished). If one has any information about previous immunizations one may be able to solve some of these mysterious deaths, but with no clues it is impossible. I have reported two cases in which anaphylaxis to pollen vaccine was the cause of death (Curry, 1962).

REFERENCES

BECKETT, A. H., TUCKER, G. T., MOFFAT, A. C. (1967). *J. Pharm. Pharmac.*, **19**, 273–294.

BIRKINSHAW, V. J., CURRY, A. S., GURD, M. R., PRICE, D. E., RANDALL, S. S., and WRIGHT, P. H. (1958). *Br. med. J.*, **2**, 463–468.

CROPPER, F. R., HEINEKEY, D. M., and WESTWELL, A. (1967). *Analyst. Lond.*, **92** 443–449.

CURRY, A. S. (1962). *Br. med. J.*, **1**, 687–689.
CURRY, A. S. (1969). *Poison Detection in Human Organs*, (2nd edn.). Springfield: Thomas.
CURRY, A. S., READ, J. F., and BROWN, C. (1969). *J. Pharm. Pharmac.*, **21**, 224–231.
CURRY, S. H. (1968). *Analyt. Chem.*, **40**, 1251–1255.
EIK-NES, K. B., and HORNING, E. C. (eds) (1968). *Gas Phase Chromatography of Steroids.* Berlin: Springer Verlag.
KANDUTSCH, A. A. (1967). *Steroids*, **10**, 31–48.
KEULEMANS, A. I. M., and CRAMERS, C. A. M. G. (1964). In *Gas Chromatography*, pp. 154–159, ed. Goldup, A. London: Institute of Petroleum.
OKERHOLM, R. A., BRECHER, P. I., and WOTIZ, H. H. (1968). *Steroids*, **12**, 435–446.
SIMMONDS, P. G., SHULMAN, G. P., and STEMBRIDGE, L. H. (1969). In *Advances in Chromatography*, pp. 230–235, ed. Zlatkis, A. Evanston, Ill.: Preston Technical Abstracts Co.
SJÖVALL, J., NYSTRÖM, E., and HAAHIT, E. (1968). In *Advances in Chromatography*, vol. 6, p. 119, ed. Giddings, J. C., and Keller, R. A. New York: Dekker.
ZLATKIS, A. (ed.) (1967). *Advances in Gas Chromatography*, Evanston, Ill.: Preston Technical Abstracts Co.
ZLATKIS, A. (ed.) (1969). *Advances in Chromatography*, Evanston, Ill.: Preston Technical Abstracts Co.

GAS CHROMATOGRAPHIC ANALYSIS OF BENZODIAZEPINES†

S. GARATTINI, F. MARCUCCI AND E. MUSSINI

Istituto di Ricerche Farmacologiche "Mario Negri", Via Eritrea, Milan, Italy

SEVERAL drugs with benzodiazepine structure show powerful pharmacological and therapeutic effects. The best-known benzodiazepines (Fig. 1) are particularly used as psychotropic drugs for the treatment of a variety of mental disturbances (Zbinden and Randall, 1967). Different subjects, however, need different doses of benzodiazepines to obtain beneficial effects and with the same dose they experience therapeutic and side effects to

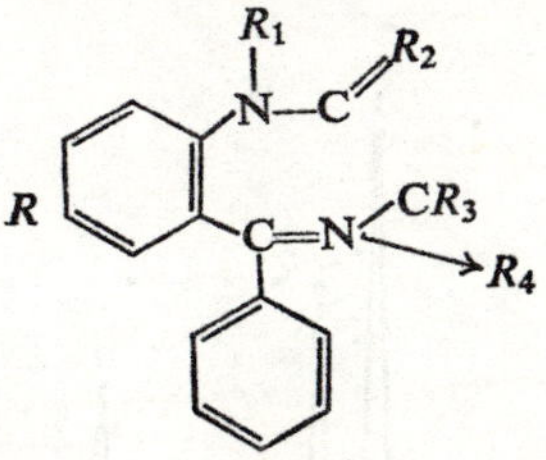

R	*R*₁	*R*₂	*R*₃	*R*₄	
Cl	CH_3	O	H_2		Diazepam
Cl	H	O	HOH		Oxazepam
NO_2	H	O	H_2		Nitrazepam
Cl		$NHCH_3$	H_2	O	Chlordiazepoxide
Cl	CH_3	H_2	H_2		Nobrium®

FIG. 1. Chemical structures of benzodiazepines used therapeutically.

different extents (Zbinden and Randall, 1967). The reason for this variability in drug activity may be related to the individual capacity to absorb, store, metabolize and dispose of a given drug (Brodie, 1964) and study of this individual variability may be a useful approach to the problem. The availability of precise and specific methods for measuring small amounts of a given drug and its metabolites in blood and tissues is important for obtaining data of these sorts.

† Paper submitted in writing after the meeting.

Gas chromatographic analysis

Gas chromatographic techniques have been particularly useful in this respect for estimating low drug concentrations. Two approaches have been followed for the analysis of benzodiazepines: (*a*) De Silva and co-workers (1964) submitted benzodiazepines to acid hydrolysis in order to obtain aminobenzophenone derivatives which are then separated by gas chromatography under suitable conditions. An important modification of this method in our experience has been the use of OV 1 as liquid phase instead of Carbowax 20 M. Carbowax 20 M is unstable above a temperature of 200° C, whereas a column packed with OV 1, which remains stable up to 300° C, can be satisfactorily used for more than one year. The sensitivity of this method

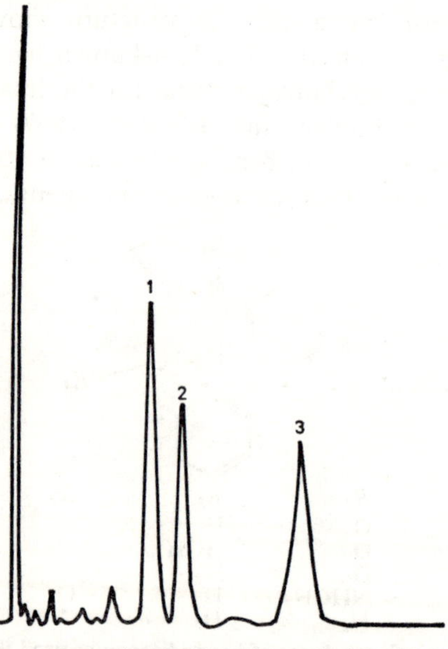

FIG. 2. Experimental conditions: *instrument*: Carlo Erba, model C; *detector*: electron capture detector (voltage: 8 V.), *column*: 2 mt. glass packed with OV 1 3% on Gas-chrom Q (60-80 mesh), temperature 195° C; *carrier gas*: nitrogen, flow rate 60 ml./min. Separation of several benzodiazepines as benzophenones: 1 = 2-amino-5-chlorobenzophenone (from chlordiazepoxide, oxazepam or *N*-desmethyldiazepam); 2 = 2-methylamino-5-chlorobenzophenone (from diazepam or *N*-methyloxazepam); 3 = *p.p′DDT* (as internal standard).

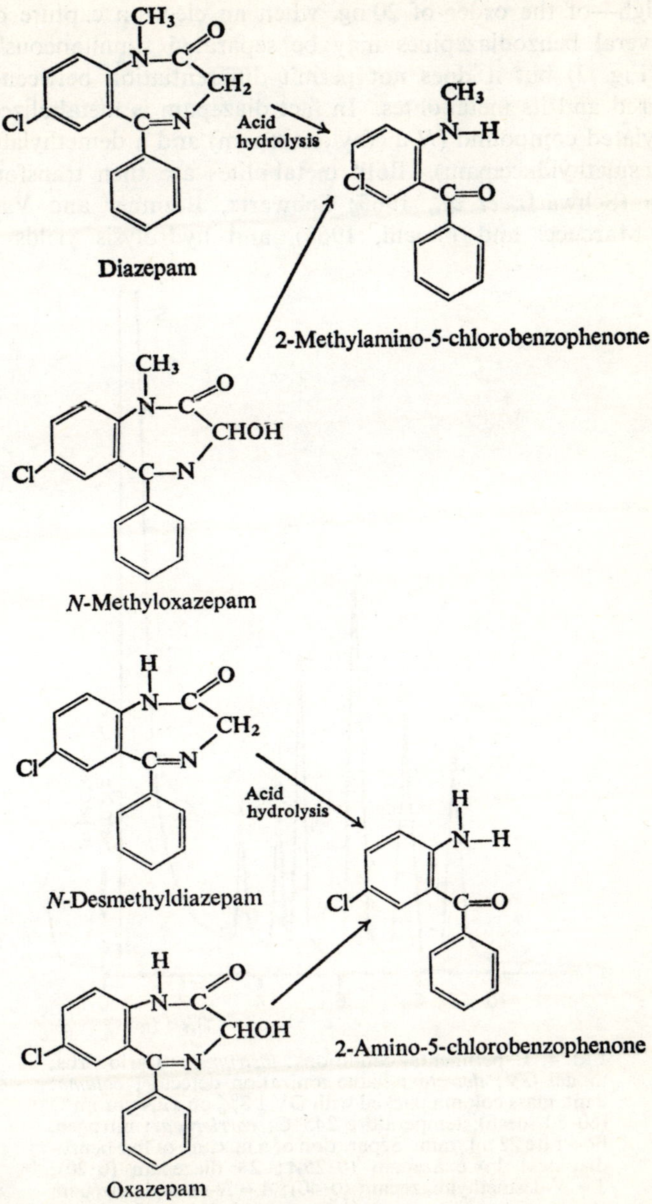

FIG. 3. Benzophenones formed by acid hydrolysis of some benzodiazepines.

is very high—of the order of 20 ng. when an electron capture detector is used. Several benzodiazepines may be separated simultaneously by this method (Fig. 2) but it does not permit differentiation between the drug administered and its metabolites. In fact diazepam is metabolized to form a hydroxylated compound (*N*-methyloxazepam) and a demethylated derivative (*N*-desmethyldiazepam). Both metabolites are then transformed into oxazepam (Schwartz *et al.*, 1965; Schwartz, Bommer and Vane, 1967; Kvetina, Marcucci and Fanelli, 1968), and hydrolysis yields the same

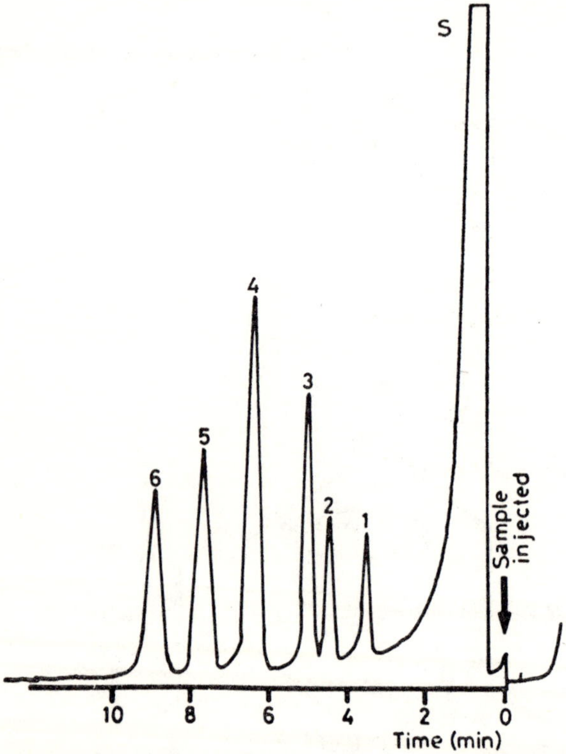

FIG. 4. Experimental conditions: *instrument*: Carlo Erba, model GV; *detector*: flame ionization detector; *column*: 2 mt. glass column packed with OV 1 3% on Gas-chrom Q (60–80 mesh), temperature 245°C; *carrier gas*: nitrogen, flow rate 22 ml./min. Separation of a mixture of five benzodiapines: 1 = oxazepam (0·25)†; 2 = diazepam (0·20); 3 = *N*-desmethyldiazepam (0·40); 4 = *N*-methyloxazepam (1·0); 5 = nitrazepam (1·0); 6 = 2-*N*-benzylamino-5-chlorobenzophenone (used as internal standard).

† Numbers in parentheses = sensitivity in microgrammes.

aminobenzophenone derivative for diazepam or *N*-methyloxazepam and for *N*-desmethyldiazepam or oxazepam (see Fig. 3); (*b*) this suggested to Marcucci, Fanelli and Mussini (1968) the need for a gas chromatographic method suitable for measuring benzodiazepines as such, without chemical manipulations. The method takes advantage of the availability of the partition liquid OV 1 which permits the separation of many benzodiazepines. The main experimental conditions are shown in Fig. 4 (flame ionization

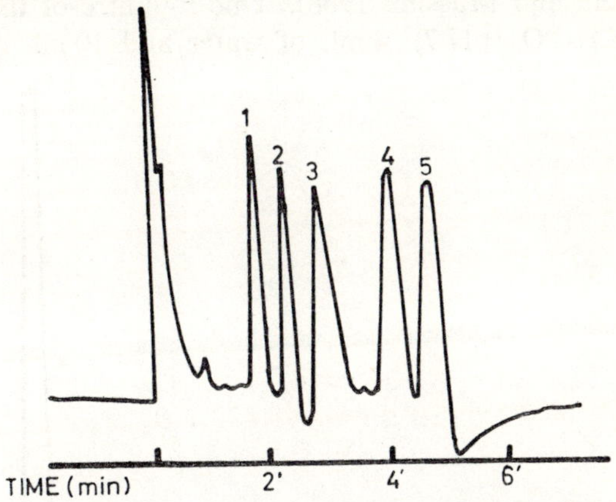

FIG. 5. Experimental conditions: *instrument*: Carlo Erba, model GI; *detector*: electron capture detector (voltage: 42 V.); *column*: 4 mt. glass column packed with OV 1 3% on Gas Chrom Q (60–80 mesh), temperature 245° C; *carrier gas*: nitrogen, flow rate 33 ml./min. Separation of a mixture of four benzodiazepines. 1 = oxazepam (2·5)†; 2 = diazepam (0·5); 3 = *N*-desmethyldiazepam (1); 4 = *N*-methyloxazapam (5); 5 = 2-*N*-benzylamino-5-chlorobenzophenone (used as internal standard).

† Numbers in parentheses = sensitivity in nanogrammes.

detection) and Fig. 5 (electron capture detection). The use of an electron capture detector has the advantage of increasing the sensitivity of the method about a hundred-fold compared with the flame ionization detector. The sensitivity is different for the various compounds. The method also allows the resolution of diazepam-N_4-oxide from *N*-desmethyldiazepam-N_4-oxide (Fig. 6). Further, nitrazepam can be separated by its reduced metabolite (Rieder, 1965; Randall *et al.*, 1965), characterized by an amino group in position 7 (aminazepam). This separation requires an increase of the column temperature from 245 to 280° C (Fig. 7). These gas chromatographic analyses can be made quantitative by the use of an internal standard, usually

2-*N*-benzylamino-5-chlorobenzophenone. When the N_4-oxide derivatives are separated, *N*-methyloxazepam is a suitable internal standard.

Extraction of benzodiazepines from biological materials

The extraction of benzodiazepines is relatively simple although less so with small concentrations of drug. De Silva and co-workers (1964) described a method that can also be used with non-hydrolysed benzodiazepines (Marcucci, Fanelli and Mussini, 1968). One millilitre of the sample, 2 ml. of buffer (1 M KH_2PO_2 : pH 7), 4 ml. of water and 10 ml. of diethyl ether

FIG. 6. Experimental conditions: *instrument*: Carlo Erba, model GV; *detector*: flame ionization detector; *column*: 2 mt. glass column packed with OV 1 3% on Gas-chrom Q (60–80 mesh), temperature 245° C; *carrier gas*: nitrogen, flow rate 22 ml./min. Separation of a mixture of two benzodiazepines: 1 = *N*-methyloxazepam (used as internal standard); 2 = diazepam-N_4-oxide (0·5)†; 3 = *N*-desmethyldiazepam-N_4-oxide (1·0).

† Numbers in parentheses = sensitivity in microgrammes.

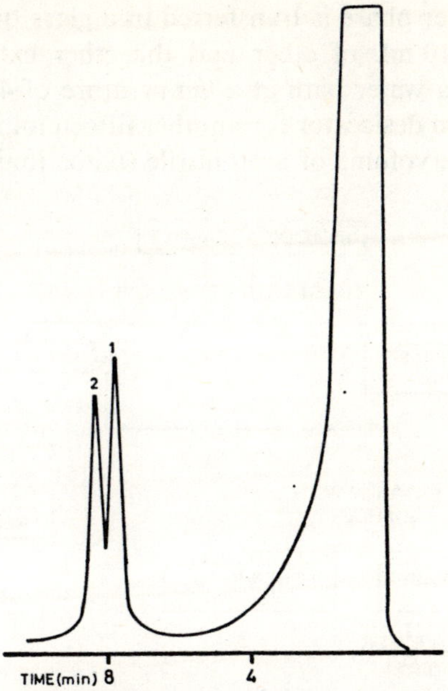

FIG. 7. Experimental conditions: *instrument*: Carlo Erba, model C; *detector*: flame ionization detector; *column*: 4 mt. glass column packed with OV 1 3% on Gas-chrom Q (60–80 mesh), temperature 280°C; *carrier gas*: nitrogen, flow rate 70 ml./min. Separation of a mixture of two benzodiazepines: 1 = nitrazepam (0·5)†; 2 = aminazepam (1·0).

† Numbers in parentheses = sensitivity in microgrammes.

(containing not more than 0·00005 per cent of peroxides—the bottle must be opened on the day that the extractions are carried out—)are mixed in a glass-stoppered centrifuge tube. The tubes are shaken on a reciprocating shaker for ten minutes and then centrifuged at a temperature of 0°C for

TABLE I

RECOVERY OF BENZODIAZEPINES FROM WATER AND BIOLOGICAL MATERIALS

	Percentage recovery ± standard error		
Drug	*water*	*blood*	*liver microsomes*
Diazepam	92 ± 3	90 ± 3	97 ± 1
N-desmethyldiazepam	100 ± 4	100 ± 3	86 ± 2
N-methyloxazepam	93 ± 1	65 ± 2	89 ± 1
Oxazepam	91 ± 1	63 ± 2	80 ± 1
Nitrazepam	96 ± 2	87 ± 3	90 ± 2

five minutes. The ether phase is transferred in a glass tube; the water phase is re-extracted with 10 ml. of ether and the ether extracts are combined and evaporated over a water bath at a temperature of 45° C. The sample is then dried in a vacuum desiccator for another fifteen minutes and the residue dissolved in a suitable volume of acetonitrile (flame ionization detection) or

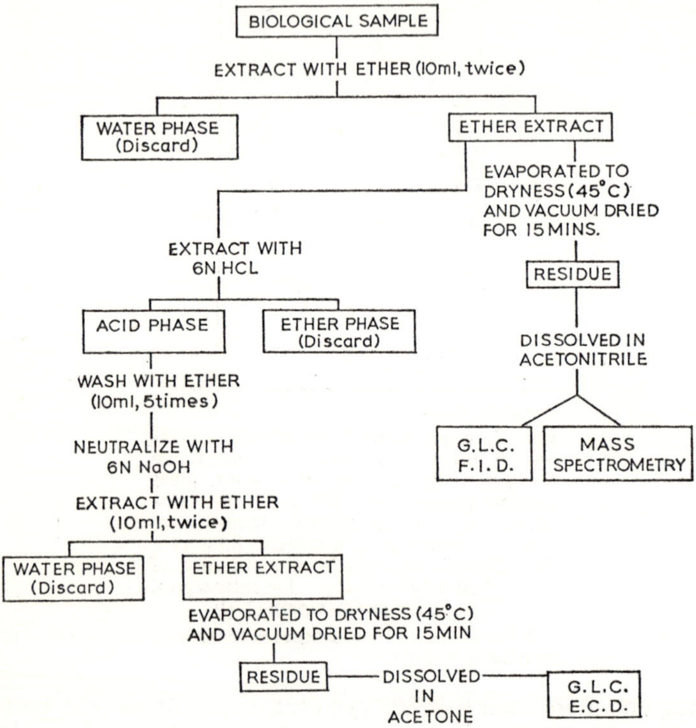

FIG. 8. Analytical procedure for the determination of benzodiazepines. G.L.C. = gas liquid chromatography; F.I.D. = flame ionization detector; E.C.D. = electron capture detector.

acetone (electron capture detection). Care is taken to ensure uniform distribution of material by tapping and stirring the tube for one minute. A suitable aliquot, from 1 to 3 μl, is then chromatographed. Table I shows the recovery of diazepam and its metabolites after extraction from water, blood or liver microsomes. Further, nitrazepam can be satisfactorily extracted from rat blood. When an electron capture detector is used, the method of extraction must be modified to allow sufficient purification of the samples. In these circumstances the recovery of the hydroxylated compounds is usually lower than the recovery reported in Table I. Fig. 8 shows diagrammatically the various steps of benzodiazepine extraction. The identification of benzodiazepines extracted from biological samples was confirmed in several instances by mass spectrometry.

Application of gas chromatographic methods to biological problems

The availability of methods for measuring benzodiazepines in blood and tissue has permitted the solution of some of the problems concerning the pharmacology of diazepam.

Diazepam has strong anticonvulsant activity (Banziger, 1965; Zbinden and Randall, 1967), but this lasts for different periods of time in various animal species. The antimetrazol activity of diazepam, for example, persists for about 3 hours in the rat and 24 hours in the mouse (Marcucci *et al.*, 1968). A possible reason for this difference may be established by measuring the level of diazepam in the brain of these two species. Using the method of De Silva and co-workers (1964), we found that the level of diazepam was similar in rats and mice but an accumulation of the *N*-desmethyl metabolites was present only in mice. The *N*-desmethyl metabolites could be separated into *N*-desmethyldiazepam and oxazepam by Marcucci and co-workers' (1968) method, previously mentioned. Preliminary results show that the sequence of the formation of these metabolites is such that in the brain of the mouse a peak of diazepam is followed by the appearance of *N*-desmethyldiazepam and finally of oxazepam. *N*-desmethyldiazepam and oxazepam show a similar degree of anticonvulsant activity, which is greater than that of diazepam (Marcucci *et al.*, 1968; Zbinden and Randall, 1967), and these biochemical findings may explain the longer-lasting anticonvulsant activity of diazepam in the mouse.

In another investigation we tried to establish the reason for the difference in the metabolism of diazepam in the rat and the mouse. Since diazepam is metabolized in the liver by the endoplasmic-reticulum system, liver from both rat and mouse was homogenized, centrifuged at $105\,000 \times g$ and the microsomal fraction, with the necessary co-factors added, (Kato and Takayanaghi, 1966) were incubated with diazepam under the experimental conditions reported in the legend for Fig. 9. It is evident from Fig. 9 that rat liver microsomes hydroxylate diazepam with the formation of *N*-methyloxazepam while the *N*-demethylation is a minor metabolic pathway. In microsomes of the mouse liver the major metabolic transformation of diazepam was the *N*-demethylation while hydroxylation was very limited. The different types of metabolic pathways in the liver of the mouse and the rat account for the different accumulation *in vivo* of the *N*-desmethyl metabolites. But although an accumulation of oxazepam occurred in the brain of the mouse *in vivo*, this could not be shown by liver microsomes *in vitro*. Preliminary studies indicate that the presence of diazepam inhibits both the further hydroxylation of *N*-desmethyldiazepam and the *N*-demethylation of *N*-methyloxazepam. But when either *N*-desmethyldiazepam or *N*-methyloxazepam instead of diazepam were added to mouse liver microsomes, oxazepam was produced.

With this experimental background it was of interest to establish the

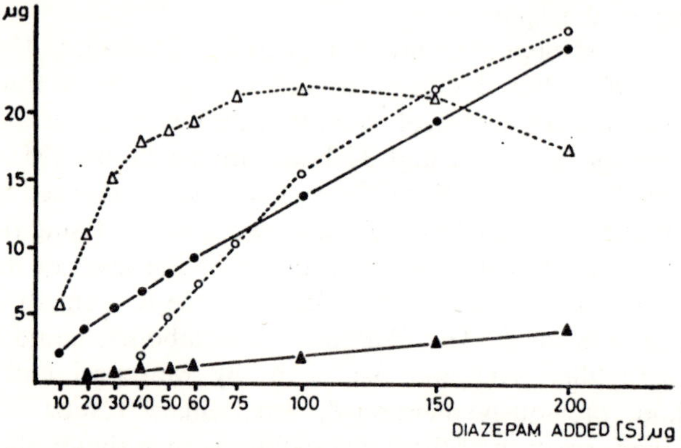

FIG. 9. *N*-desmethyldiazepam (△---△) and *N*-methyloxazepam (○---○) produced from diazepam by microsomal system in the mouse. *N*-desmethyldiazepam (▲——▲) and *N*-methyloxazepam (●——●) produced from diazepam by microsomal system in the rat. Experimental conditions: regularly increasing doses of diazepam were incubated with 2·5 ml. of microsomal suspensions containing 24 mg. of protein ≡ 1 g. of fresh liver. The incubation mixture consisted of NADP (1·5 µM.), glucose-6-phosphate (50 µM.), glucose-6-phosphate dehydrogenase (0·5 units), magnesium chloride (25 µM.), nicotinamide (50 µM.); 1·4 ml. of 0·2M. phosphate buffer (pH 7·4); 0·45 ml. 1·15% (w/v) KCl; and water to a final volume of 5 ml. The mixtures were incubated for 10 min. under air at a temperature of 37°C in a Dubnoff metabolic shaking incubator. Each point on the graph represents the average of four experiments.

presence of *N*-desmethyl metabolites in human beings treated with single doses of diazepam. Previous studies have indicated that diazepam can be measured in the blood of patients and that for a period of about 5 hours after ingestion *N*-desmethyl metabolites are not formed (De Silva, Koechlin and Bader, 1966; Garattini, 1969). However, in patients under continuous treatment with diazepam (5 mg. orally three times per day) *N*-desmethyldiazepam appeared in the blood after two days. These findings agree with those of De Silva, Koechlin and Bader (1966).

Table II shows that during diazepam therapy the concentrations of *N*-desmethyl metabolites in blood may be higher than the concentration of diazepam.

It is remarkable that subjects receiving the same dose of diazepam show a large variation in blood levels of the drug. Garattini (1969) has shown a correlation between the peak level of diazepam in the blood and the presence of such side effects as drowsiness.

TABLE II

BLOOD LEVELS OF DIAZEPAM AND ITS *N*-DESMETHYLATED METABOLITES IN MAN AFTER CONTINUOUS TREATMENT WITH DIAZEPAM†

	Blood levels‡	
Patients	*Diazepam* (μg/ml.)	*N-desmethylated* (*metabolites* μg/ml.)
A	0·123	0·184
B	0·243	0·314
C	0·152	0·243
D	0·180	0·247
E	0·104	0·120
F	0·194	0·226
G	0·106	0·144
H	0·231	0·281

† 5 mg., three times per day orally, for 48 hours.
‡ 8 hours after last administration of drug.

These findings are offered as an exemplification of the great value of gas chromatographic techniques in the study and interpretation of the pharmacological and therapeutic activity of the benzodiazepines.

SUMMARY

Two gas chromatographic techniques are available for the measurement of several of the benzodiazepines in biological materials. The experimental conditions, limitations and sensitivity of the methods are summarized and discussed. Examples of the importance of gas chromatographic techniques in the understanding of some pharmacological and therapeutic problems regarding the benzodiazepines are presented.

Acknowledgements

The experimental work reported in this paper has been supported by the National Institutes of Health, Bethesda, Maryland, USA, contract N. DHEW/PHS/NIH, PH 43-67-83.

We are grateful to Professor G. C. Jommi, Department of Organic Chemistry, University of Milan, for the synthesis and purification of several benzodiazepines, and to Mr. R. Fanelli for his skilful technical help.

REFERENCES

Banziger, R. F. (1965). *Archs int. Pharmacodyn. Thér.*, **154**, 131–136.

Brodie, B. B. (1964). In *Absorption and Distribution of Drugs*, pp. 199–251, ed. Binns, T. B. Edinburgh: Livingstone.

De Silva, J. A. F., Koechlin, B. A., and Bader, G. (1966). *J. pharm. Sci.*, **55**, 692–702.

De Silva, J. A. F., Schwartz, M. A., Stefanovic, V., Kaplan, J., and D'Arconte, L. (1964). *Analyt. Chem.*, **36**, 2099–2105.

Garattini, S. (1969). In *Proc. IV Collegium Internationale Neuropsychopharmacologicum Meeting* (held in Tarragona, Spain, 1968), in press. Amsterdam: Excerpta Medica.

Kato, R., and Takayanaghi, M. (1966). *Jap. J. Pharmac.*, **16**, 380–390.

Kvetina, J., Marcucci, F., and Fanelli, R. (1968). *J. Pharm. Pharmac.*, **20**, 807–808.

Marcucci, F., Fanelli, R., and Mussini, E. (1968). *J. Chromat.*, **37**, 318–320.

Marcucci, F., Guaitani, A., Kvetina, J., Mussini, E., and Garattini, S. (1968). *Europ. J. Pharmac.*, **4**, 467–470.

Randall, L. O., Schallek, W., Scheckel, C., Bagdom, R. E., and Rieder, J. (1965). *Schweiz. med. Wschr.*, **95**, 334–337.

Rieder, J. (1965). *Arzneimittel-Forsch.*, **15**, 1134–1148.

Schwartz, M. A., Bommer, P., and Vane, F. M. (1967). *Archs Biochem. Biophys.* **121**, 508–516.

Schwartz, M. A., Koechlin, B. A., Postma, E., Palmer, S., and Krol, G. (1965). *J. Pharmac. exp. Ther.*, **149**, 423–435.

Zbinden, G., and Randall, L. O. (1967). *Adv. Pharmacol.*, **5**, 231–291.

GAS CHROMATOGRAPHIC AND SPECTROMETRIC TECHNIQUES

R. P. W. Scott

Unilever Research Laboratory, Colworth House, Sharnbrook, Bedford

Gas chromatography is primarily a separation technique but with the aid of a linear detector the technique can provide quantitative analysis of volatile mixtures. Under well controlled conditions the measurement of retention ratios can permit the identification of substances provided reference data for these substances are available. If the substance is completely unknown or reference retention data are unavailable gas chromatography is of little help in the identification of components and it is necessary to use other ancillary techniques. The most useful ancillary technique to gas chromatography is mass spectroscopy, as in its modern form it is fast enough to obtain spectra during the elution of the peak and sensitive enough to be able to operate on the range of sample loads normally used with the gas chromatograph. A typical fast-scan mass spectrometer that can be used on-line with a gas chromatograph is the AEI MS 12 mass spectrometer. A fast-scan instrument should have scan times down to 1 sec. and a mass range of up to 800. Although most mass spectrometers can provide spectra from submicrogramme quantities of sample the concentration of sample entering the inlet system must be adequate. Due to the band dispersion that occurs in a chromatograph the concentration of the eluent, in the carrier gas, even at the peak maximum may be insufficient to provide good mass spectra and it is necessary to concentrate the eluent in the carrier gas before entering the spectrometer inlet system.

Several methods have been devised for concentrating column eluents. One of the first to be described was that devised by Ryhage (1964) which consisted of an alignment of jets and venturi placed between the column exit and the mass spectrometer. Helium was used as the carrier gas and, while transversing between each jet and its complementing venturi, helium diffused away preferentially resulting in an enrichment of the solute vapour. The selective diffusion of solute through a thin silicone rubber membrane is another method of solute enrichment devised by Llewellyn (1966). The carrier gas passes over the membrane and the vapour that diffuses through the membrane passes to the mass spectrometer inlet. The method devised by Lipsky, Horvath and McMurray (1966) involved the connexion of the column directly to the mass spectrometer via a Teflon capillary tube. This tube was heated to a temperature of about 270° C, at which temperature the carrier gas, helium, diffuses through the walls whereas the solute vapour does not, resulting in an increase in vapour concentration within the Teflon tube. One of the simplest and most useful methods of solute concentration is that developed by Watson and Biemann (1964) (see Fig. 1). It consists of a jacketed tube of sintered glass through which the carrier gas passes. The tube is heated to a temperature of about 200° C and the jacket maintained at reduced pressure. The helium diffuses rapidly through the sintered tube into the jacket whereas the solute vapour diffuses only very slowly. The resulting concentrated vapour then passes out of the sintered tube and into the mass spectrometer. A photograph of the Biemann concentrator fitted to the MS 9 spectrometer is shown in Fig. 2.

Operating the mass spectrometer at a scan time of 1 sec. or less allows the different parts of a peak to be monitored and, if two eluents are incompletely

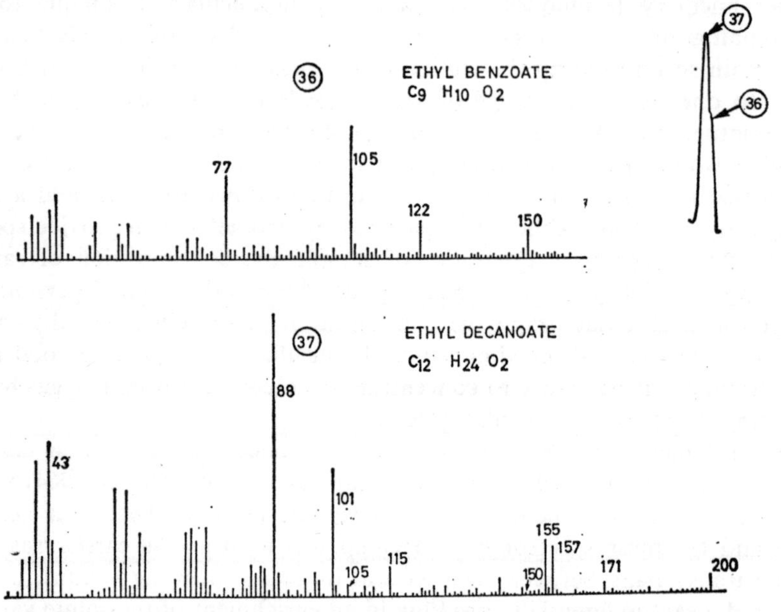

FIG. 3. Mass spectra of ethyl benzoate and ethyl decanoate taken from an unresolved peak.

FIG. 1. The beam and concentrator (see text).

FIG. 2. The beam and concentrator fitted to the MS 9 spectrometer.

[*to face page 174*

resolved, by monitoring each side of the component peak uncontaminated spectra of each substance can be obtained. An example of the procedure is seen in Fig. 3 where clean spectra of ethyl benzoate and ethyl decanoate are shown although the two substances have not been resolved by the chromatograph.

The spectra produced by a mass spectrometer can be recorded as an analogue signal on magnetic tape, which then, after passing through an analogue to digital convertor, can be fed direct to a computer and used for sorting and identification.

The success of this procedure will depend, of course, on the reference spectra available. Table I shows the results of a sorting procedure by an IBM 7030 computer. The reference data contained a large number of

TABLE I

RESULTS FROM A SORTING PROCEDURE BY THE IBM 7030 COMPUTER

Method 4	*Compound*	*Probability*
	Alpha-terpineol	0·603
	Alpha-terpineol	0·597
	Limonene	0·592
	Alpha-terpineol	0·583
	Dipentene	0·574
	Dipentene	0·518
	Beta-pinene	0·481
	trans-Caran-*cis*-3-ol	0·476
	cis-Caran-*trans*-3-ol	0·476
	D-Limonene	0·476
	1-Methyl-5-ethylidene cycloheptene	0·462

spectra obtained from different mass spectrometers, and it can be seen that the unknown is identified with a high degree of probability and that all the spectra for the unknown α–terpineol—were selected as likely comparison spectra.

The fast-scan, low-resolution mass spectrometer has considerable limitations, however, when reference spectra are unavailable and one is dealing with a completely unknown substance. The high resolution instrument will measure the mass of the parent molecular ion with an accuracy of a few ppm and thus permit the calculation of the molecular formula. A modern, high resolution spectrometer can scan a spectrum in 10 sec. at a high enough resolution (10 000) to permit the masses of the ions to be measured with an accuracy of better than 10 ppm, needing less than 1 μg. of sample.

Due to the relatively long scan required this spectrometer is best operated with packed columns that have longer analysis times. Again, the spectra can be recorded on magnetic tape, digitized and fed to the computer. The

TABLE II
ELEMENT MAP DERIVED FROM A COMPUTER OF A HIGH RESOLUTION MASS SPECTRUM

METHYL STEARATE SAMPLE. SAMPLE SIZE...........25 NG.

-----

1 1

CALCULATED MASS	ERR	C12/13	H	N	O	MEASURED MASS	NO. PTS	INTENSITY
299.2949	-2.07	19/0	39	0	2	299.2928	16	+++++++
299.2904	2.38	18/1	38	0	2			
298.2870	.85	19/0	38	0	2	298.2879	24	+++++++++
269.2435	.73	16/1	32	0	2	269.2443	10	++++++
267.2687	.42	18/0	35	0	1	267.2691	16	+++++++
256.2357	.00	15/1	31	0	2	256.2357	8	+++++
255.2323	- .18	16/0	31	0	2	255.2321	20	++++++++
241.2167	1.19	15/0	29	0	2	241.2179	11	++++++
NO COMP CALC						223.0968	20	++++++++
213.1853	-1.22	13/0	25	0	2	213.1842	14	+++++++
NO COMP CALC						205.0856	12	++++++
200.1775	-1.55	12/0	24	0	2	200.1760	8	+++++
199.1697	- .45	12/0	23	0	2	199.1693	18	++++++++
185.1541	.67	11/0	21	0	2	185.1547	14	+++++++
NO COMP CALC						182.0927	7	+++++
NO COMP CALC						167.0333	17	+++++++
157.1227	- .24	9/0	17	0	2	157.1225	15	++++++
NO COMP CALC						150.0276	20	++++++++
NO COMP CALC						149.0240	34	+++++++++++
143.1071	.12	8/0	15	0	2	143.1072	24	+++++++++
130.0993	.61	7/0	14	0	2	130.0999	8	+++++
129.0915	.64	7/0	13	0	2	129.0921	19	+++++++
123.1173	.54	9/0	15	0	0	123.1179	7	+++++
122.0731	- .51	8/0	10	0	1	122.0726	17	+++++++
122.0367	.71	7/0	6	0	2	122.0374	7	+++++

computer can then provide the elemental compositions of all the fragments and the parent molecular ion.

An example of a listing of composition of ions derived by a computer from an analogue tape of high resolution mass spectra is shown in Table II.

The gas chromatography–mass spectrometry combination has considerable application to the biological and medical fields of research. An example of the use of the system to monitor blood lipids is shown in Fig. 4. The chromatogram shows the fatty acid and cholesterol methyl esters. The

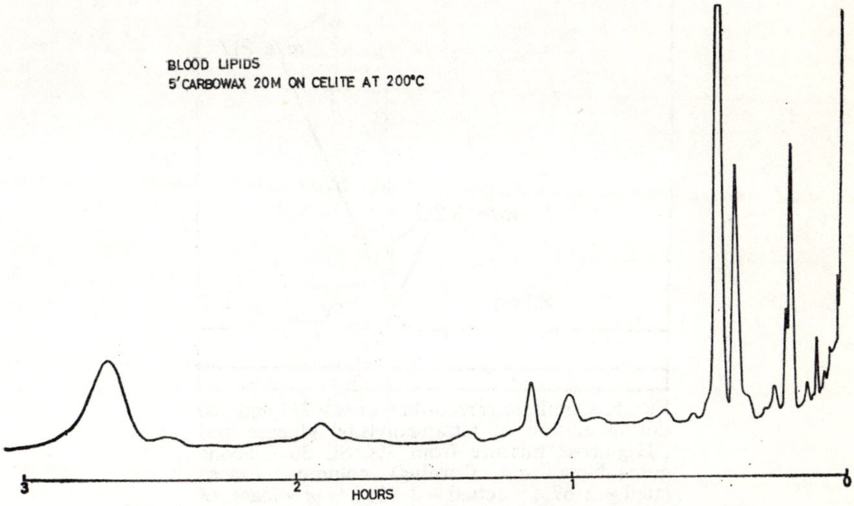

FIG. 4. Chromatogram of blood lipids identified from mass spectra.

carbon number of each fatty acid can be immediately obtained and numbered, and the cholesterol ester easily identified. Another interesting application of this system is to the technique in which metabolic pathways are followed by means of deuterium-labelled compounds. It is very difficult to separate hydrogen and deuterium isotopes of organic compounds by gas chromatography, though some degree of separation usually occurs, but by using the mass spectrometer as a detecting system in the way developed by Sweeley and co-workers (1966) the quantitative determination of each isotope is simple. The mass spectrometer is made to detect, successively, during the elution of the peak, ions characteristic of the isotopically labelled compound which, in the example given, have masses of 217 (220 for the isotopes of trimethyl silyl derivatives of glucose). This is achieved by pulsing the accelerating voltage by a square wave, the amplitude of which is adjusted to that required to change from an m/e of 217 to an m/e of 220.† The resulting peak profile is shown in Fig. 5. It can be seen that two distinct peaks are produced for the two isotopic species although, from a chromatographic point of view, the two peaks are virtually coincident.

The mass spectrometer–gas chromatograph combination has also been used to examine the Krebs cycle acid methyl esters, a chromatogram of which is shown in Fig. 6. The molecular weight of each eluent has been determined by the mass spectrometer during chromatographic development.

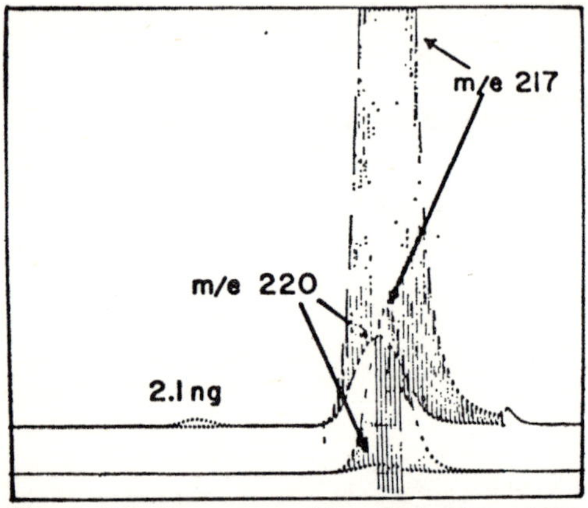

FIG. 5. Continuous recording of m/e 217 and 220 during elution of tetramethylsilyl glucose and [^{2}H]glucose mixture from 3% SE 30 (silicone gum—Dow and Corning) column. Calculated = 2·67%; actual = 2·71%. m = mass of particle; e = charge of particle.

During the toxological studies recently carried out on food additives in the Unilever Laboratories, animal testing had been satisfactorily completed and the stage reached at which human volunteers were being used. The metabolism of the food additives was followed by chromatographing ether extracts of alkaline urine. The chromatogram of a sample of the urine is shown in Fig. 7. It became apparent that in certain subjects an odd peak developed which had not been noted during animal testing. The last peak was examined on a gas chromatography–mass spectrometry apparatus and the abnormal metabolite was identified as nicotine. It was found also that only people who were smokers excreted nicotine. The quantity excreted increased with the degree of smoking.

The use of infra-red (i.r.) spectroscopy and nuclear magnetic resonance (n.m.r.) spectroscopy has a somewhat more limited application to biological

† m = mass of particle; e = charge of particle.

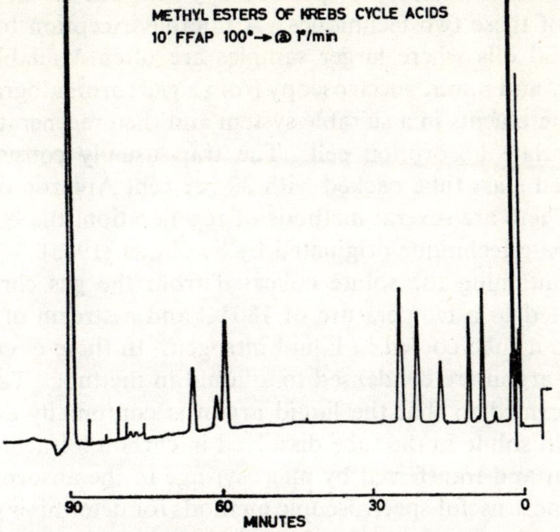

FIG. 6. Chromatogram of the methyl esters of Krebs cycle acids identified by mass spectra. 3·048 m. (10 ft.) column carrying FFAP (Airograph), programmed from 100–200°C at 1°/min.

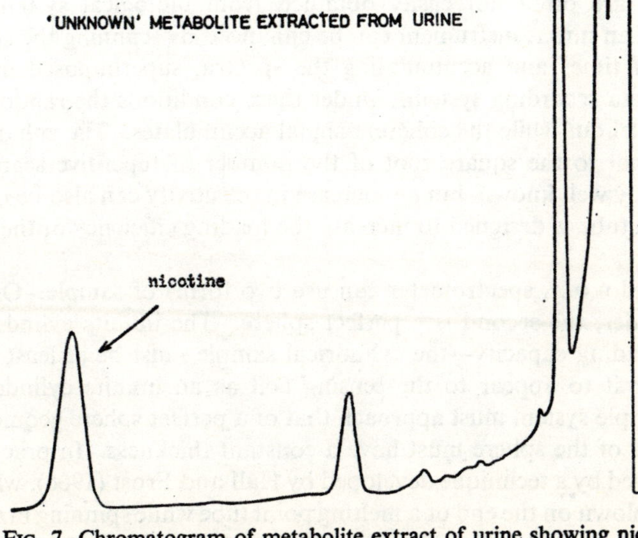

FIG. 7. Chromatogram of metabolite extract of urine showing nicotine identified from mass spectra.

and medical research than mass spectroscopy has, due to the larger sample requirements of these two techniques. A notable exception to this is in the field of essential oils where larger samples are often available. To obtain samples for i.r. and n.m.r. spectroscopy from a gas chromatograph it is necessary to trap the eluents in a suitable system and then regenerate and transfer to the appropriate absorption cell. The trap usually consists of a short length of cooled glass tube packed with 30 per cent Apiezon on 100–120 BS mesh celite. There are several methods of regeneration, the best of which is the total trapping technique originated by Swoboda (1963).

The trap containing the solute collected from the gas chromatographic column is heated to a temperature of 180° C and a stream of argon passed through it into a tube cooled in liquid nitrogen. In these circumstances the solute and the argon are condensed to a liquid in the tube. The tube is then very slowly warmed so that the liquid argon is continually evaporated off, and the residual solute in the tube dissolved in carbon tetrachloride or other suitable solvent and transferred by microsyringe to the absorption cell.

One of the most useful spectroscopic methods for determining the structure of complex organic material is n.m.r. spectroscopy. This technique, however, normally requires 2–3 mg. of an individual component to obtain a satisfactory spectrum. Such quantities impose an almost impossible task on high efficiency columns if the minor components of a mixture are to be identified.

Further, such quantities of material in pure form suitable for spectroscopic examination are often not easily obtained from biological systems. The sensitivity of an n.m.r. instrument can be enhanced by scanning the spectrum a number of times and accumulating the spectra, superimposed in digital form on a data recording system. Under these conditions the random noise tends to cancel out while the coherent signal accumulates. The enhancement is proportional to the square root of the number of repetitive scans. This system is fairly well known, but an increase in sensitivity can also be obtained if the sample tube is designed to increase the loading efficiency of the sensing coil.

The normal n.m.r. spectrometer can use two forms of sample. One is an infinite cylinder, the second is a perfect sphere. The infinite cylinder has a very poor loading capacity—the cylindrical sample must be at least 40 mm. in length for it to appear to the sensing coil as an infinite cylinder. The spherical sample system must approach that of a perfect sphere geometrically and the walls of the sphere must have a constant thickness. In practice this can be achieved by a technique developed by Hall and Frost (1966), where the microcell is blown on the end of a melting point tube while spinning in a chuck. With this system an excellent spectrum can be obtained from 25 μg. of ethyl benzene using about 100 accumulated scans and, generally, good spectra can be obtained from 150–200 μg. of sample. When using the glass spherical cell with trapped eluents from a gas chromatograph, the sample is regenerated directly into the cell by immersing it in liquid nitrogen and subsequently evaporating off the argon in the usual way.

The combination of gas chromatography and spectrometry is a very powerful system for identifying unknown volatile materials and has a wide scope of application in biological and medical research. At present, liquid chromatography is developing towards the same degree of sophistication as gas chromatography and it is probable that in the not too distant future liquid chromatographs will be directly linked with either i.r. or u.v. spectrometers or even, perhaps, with n.m.r. spectrometers.

SUMMARY

The design of solute concentration systems for use with the gas chromatograph–mass spectrometer combination are described and the automatic data handling of mass spectra by computers is discussed. Examples of the application of the mass spectrometer–gas chromatographic system to biological problems are also given. The design of apparatus for the normal trapping of column eluents for subsequent examination by infra-red and nuclear magnetic resonance spectrometry are considered and the use of microcells in nuclear magnetic resonance spectroscopy to enhance the sensitivity is discussed.

REFERENCES

HALL, G., and FROST, D. (1966). *Molec. Phys.*, **10,** 191–200.

LIPSKY, S. R., HORVATH, C. G., and MCMURRAY, W. J. (1966). *Analyt. Chem.*, **38,** 1585–1587.

LLEWELLYN, P. (1966). Presented to Combined GC/MS Technicians Symposium at *Pittsb. Conf. analyt. Chem. appl. Spectrosc.* [Pittsburg, Feb., 1966]. Offprint, circulated privately.

RYHAGE, R. (1964). *Analyt. Chem.*, **36,** 759–764.

SWEELEY, G. C., ELLIOTT, W. H., FRIES, I., and RYHAGE, R. (1966). *Analyt. Chem.*, **38,** 1549–1553.

SWOBODA, P. A. T. (1963). *Nature, Lond.*, **199,** 31–32.

WATSON, J. T., and BIEMANN, K. (1964). *Analyt. Chem.*, **36,** 1135–1137.

DISCUSSION

Warren: How portable are the mass spectrometers you have described? I appreciate that the size will depend on the degree of resolution.

Scott: The AEI high resolution mass spectrometer MS 9 occupies a space of about $3 \cdot 5 \times 2 \cdot 5 \times 1 \cdot 5$ m. together with a console which occupies a space of approximately $2 \times 1 \times 1$ m. It is also very heavy. The small quadrupole mass spectrometer described by Professor Lipsky, however, could be developed in a conveniently portable form.

Warren: What sort of mass spectrometer is used in spacecraft?

Scott: A small quadrupole mass spectrometer is combined with the gas chromatograph.

Lipsky: The total weight of this instrument is about 6 kg.

Warren: What is the resolution?

Scott: It has a mass range of 400.

Purnell: A similar quadrupole instrument is now commercially available in the UK; it costs about £1000 and has a resolution of about 150.

Lipsky: The analysers in these small instruments are residual gas analysers, and are differently designed from those used in the large mass spectrometers. Analyser heads that are rejected because of malalignment of the rods to very small tolerances are used as residual gas analysers; analyser heads that are perfectly aligned are incorporated in instruments used for organic analysis.

Purnell: Is it because of the computer of average transients (the data accumulating system) in your machine that you can manage with such small amounts of material, Dr. Scott?

Scott: Partly because of the data accumulating system associated with the n.m.r. spectrometer and partly because of the microcell used in our apparatus (see my paper, this volume, p. 173).

Lipsky: If we could use the Fourier transform analysis to enhance the signal (see Ernst and Anderson, 1966), analysis time could be decreased by a factor of 100. A run taking 10 minutes in Dr. Scott's apparatus would now only take 6 seconds. This would enable us to achieve an additional one or two orders of sensitivity.

Scott: There is also room for imaginative development of the n.m.r. spectrometer.

Lipsky: We are using a recently developed n.m.r. system which has a signal to noise ratio of 80:1 and an improved magnet. This is an excellent system; it avoids extraneous side bands.

Scott: We have now reached the stage at which enough material could be provided by a chromatograph to be studied by n.m.r. technique.

Payne: Would you define n.m.r. for the members of the group who, like myself, do not know much about it, Dr. Scott?

Scott: Nuclear magnetic resonance (n.m.r.) is by far the most powerful spectrometric technique available. Any spinning nucleus which either has an odd charge or is asymmetric in form will constitute a circular current and thus behave as a small magnet. If such a nucleus is placed in a magnetic field the axis of rotation will tend to be aligned with the field and thus, according to Newton's law, the nucleus will precess. The precession frequency will depend on the strength of the external magnetic field and on the magnetic field contributed by neighbouring atoms. Thus the precession frequency will be directly related to the nature of the neighbouring atoms from their characteristic contributions to the magnetic field. From the Quantum theory we know that there are two possible energy levels for a precessing nucleus; by subjecting the nucleus to an electromagnetic wave of the same frequency as the precessing nucleus it will absorb energy and this energy absorption can be detected. Thus, by either sweeping the field or the frequency of the electromagnetic wave, a series of absorption peaks will be obtained that describe the atomic environment of each nucleus. In proton n.m.r. spectroscopy, the atoms that are associated with each hydrogen atom are thus identified and the molecular structure of the substance examined can be obtained by suitable interpretation.

Lipsky: One can also measure other nuclei with the n.m.r. spectrometer.

Scott: That is true. I described the n.m.r. system for protons. One can use any nucleus that has an asymmetric charge or is in itself asymmetric. Phosphorus is a good example.

Purnell: N.m.r. spectrometry is an excellent specific method for the fingerprinting of organic compounds. Today's organic chemist no longer analyses chemicals in test-tubes, he commonly puts his sample (which may be of a very large molecule indeed) in this machine and within minutes he is given the structures of his unknowns. The comparison of n.m.r. and chemical methods for the identification of organic compounds is remarkably favourable to n.m.r.

Brooks: The method does have some drawbacks. Compounds like the steroids contain methylenic protons, many of which are adjacent and similar in character. Thus only part of such molecules can be determined directly by n.m.r. spectrometry; with other steroids one may need to use pattern comparisons as well. In general, this technique is not so useful for complete structure identification unless a great variety of protons or other nuclei can be examined so that the environment of each one can be completely distinguished.

Scott: But would you not agree with me that n.m.r. is the most powerful spectrometric technique available at present?

Brooks: In principle, yes.

Lipsky: And in practice; there is no doubt about it.

Martin: Siegbahn and co-workers (1967) in Sweden are developing a new technique: photoelectron spectrometry. Will this be useful in this field?

Purnell: It is too early to answer that question. Some scientists in our department are now working on a project using photoelectron spectrometry for the analysis of fungicides and pesticides. It seems likely that extremely complex molecules like these can be readily identified by this technique.

Martin: Do photoelectron and n.m.r. spectrometry both give the same sort of information?

Purnell: Yes, in that they both tell us a great deal about the internal structure of the molecule in question.

Martin: If both these techniques demonstrate, to some extent, the position of neighbouring atoms, the photoelectron spectrometer, which operates with microgramme to nanogramme amounts of material, fits better than n.m.r. with chromatographic techniques.

Purnell: Photoelectron spectrometry will ultimately play an important role in any research in which the identification of large molecules is of primary interest.

Lipsky: Most of the work with these instruments so far is unpublished and has, unfortunately, been with inorganic compounds.

Purnell: Turner's (1968) method of using a stream of excited helium atoms to eliminate the filament in a mass spectrometer is another interesting recent development; the molecules are ionized by photons corresponding to about 21 eV.

Scott: Is this a method for delivering quantified energy for ions, the molecule receiving the same amount of energy at each delivery?

Purnell: Twenty-one eV is nearly 2000 kJ (500 kcal)/mole, corresponding to a wavelength of about 500 Å. The wavelength can be increased by the use of other rare gases.

Brooks: Dr. Scott, I am worried about your illustration of a computer search for identification because of the choice of the first two compounds, α-terpineol and limonene (Table I, p. 175). These two compounds differ in every chemical and physical property (including molecular weight) except their behaviour on mass spectrometry because, in the mass spectrometer, α-terpineol is dehydrated to limonene.

Scott: This is exactly the reason we chose terpenes for the study: because they are so difficult to identify spectrometrically. If we could develop a programme to identify materials such as these the system could be shown to be efficient. But we are not advocating that this technique is the best way of identifying terpineol. Terpenes are a group of compounds that are ideal for testing out the sorting system because of the many terpene spectra available in the reference catalogue and also because many different spectrometers have been used to obtain these spectra.

Brooks: You were deliberately loading the dice against yourself.

Scott: Yes.

Brooks: Would you use this method in practice?

Scott: Oh no; we would identify terpineol from its retention ratio or retention index. We do, however, use the system generally for terpenes.

Brooks: In my view it is unreasonable to complicate the problem of identification by using an inadequate sample introduction system. If cholesterol is injected effectively into the mass spectrometer its spectrum, normally including the molecular ion, will appear (Gohlke, 1963). But if cholesterol is introduced in such a way that it is dehydrated, obviously one will not get the same spectrum. The same proposition applies to the differentiation of terpenes from terpene alcohols. They are distinguishable in many simple and cheap ways other than mass spectrometry.

Scott: This was not at all the point of the illustration. We devised an especially difficult sorting problem in order to find out if the system functions under adverse conditions. Of course we would not use this system if we had other corroborating evidence. I am not familiar with cholesterol but the practical problems must be similar to the ones we meet with the terpenes. The results I showed convinced us we had identified a substance satisfactorily in spite of there being no parent peak and despite its giving very similar spectra to those of other substances. But, I repeat, no one in their senses would look for α-terpineol by mass spectra.

Purnell: In what percentage of cases are you unable to find the parent peak? Surely with some compounds this peak must be unobtainable.

Scott: About 25 per cent of essential oils show no parent peak on the mass spectrometer. We choose a number of major peaks and then sort on the basis of their mass and relative peak heights.

Lipsky: Industrial research has failed badly in this area. If it were possible to devise an instrument with a dual source of ionization—a field source as well as the conventional source—one would always get a molecular ion. Although one would not get good fragmentation patterns in this way, this is a useful development that has so far been neglected.

Scott: Voltage scanning will produce a smaller molecular-ion peak than magnetic scanning. But magnetic scanning is significantly slower than voltage scanning. The spectra I showed were voltage scanned. Magnetic scanning is essential if one needs to see the molecular ion from a compound of high molecular weight in a low resolution mass spectrometer. Magnetic scanning gives rise to a threefold increase in scanning time.

REFERENCES

ERNST, R. R., and ANDERSON, W. H. (1966). *J. scient. Instrum.*, **37**, 93–102.

GOHLKE, R. S. (1963). *Chemy Ind.*, 946–948.

SIEGBAHN, K., and ten co-workers. (1967). *Nova Acta Regiae Societatis Scientiarum Upsaliensis*, ser. IV, **20**, 1–282. (Presentation to Royal Society of Sciences of Uppsala, December, 1965).

TURNER, D. W. (1968). *Chem. Britain*, **4**, 435.

DIGITAL COMPUTERS AND THE ANALYSIS OF CHROMATOGRAPHIC DATA

G. B. MARSON

Digital Systems Division, Kent Instruments Ltd, Luton, Bedfordshire

THE combination of chromatographic analysers with an on-line digital computer can produce significant advantages in the usefulness of the analysis. Computers have already been used quite extensively for accurate area computation and derivation of normalized milar percentages of the components in the sample. Reductions in analysis time, by the use of shorter columns for example, may reduce the separation of the peaks in the chromatograms, and the computer can employ various strategies to resolve and reallocate the peak areas.

In some applications the analyser and computer can be used as part of a control system in which gas flows or experimental conditions are varied automatically until the analysis shows the desired result.

Integration

The proportion of each component in a sample is represented by the time integral of the detector output during the peak produced by that component. The necessary integration can be done manually, by use of a planimeter, or more usefully by a mechanical disc integrator attached to a recorder. A semi-automatic system is provided by feeding the detector output to either a digital integrator or a digital computer. In either case, the main difficulty in devising a suitable input system is the need to integrate accurately peaks whose heights can differ by a factor of up to 10 000 to 1. This, in fact, calls for a dynamic range of $10^6:1$ in the input amplifier. Since the dynamic range of most commercially available amplifiers is generally below $10^4:1$, some form of automatic range-switching is normally required.

A further requirement of the input system is that of sampling the detector output sufficiently often to ensure that its shape is correctly re-created and integrated. An adequate rate is approximately five samples per second for peaks from a 6·4-mm. packed column, or up to twenty per second for a capillary column. This presents no great problem for a digital integrator serving a single chromatograph, but may limit to five or ten the number of chromatograms that can be handled simultaneously by a shared on-line digital computer. Short-term integration can, however, be performed very accurately by a purely analogue device, and the hybrid technique shown in Fig. 1 is a useful means of employing an analogue input system to enhance both the dynamic range and effective sampling frequency of an on-line computer.

In this system the chromatograph signal is integrated with respect to time on a Miller integrator. Every 25 msec, the integrator output is examined by

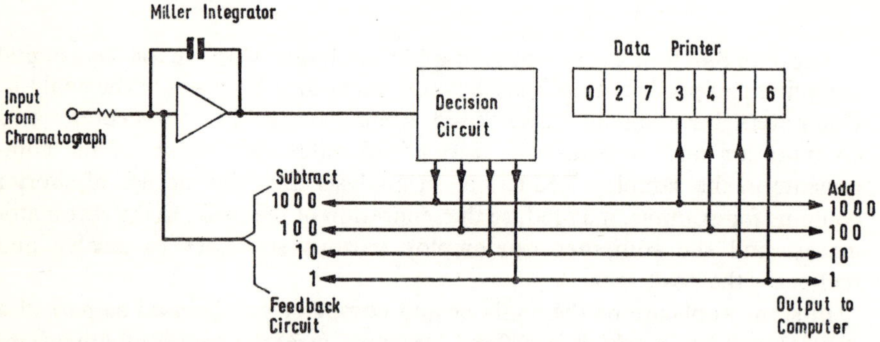

FIG. 1. Wide dynamic range integrator used as computer input device.

a decision circuit to decide whether it represents at least 1, 10, 100 or 1000 units of area. One of these values is then subtracted automatically from the integral, by a feedback circuit, and a digital signal given to the computer to increment a location, corresponding to the peak integral, by 1, 10, 100 or 1000. Using this technique an effective count rate of 40 000/sec. is achieved, and the computer is relieved of the burden of making many analogue measurements, with associated range-changing, every second.

This technique would also reduce the amount of data to be stored for transfer to an off-line computer for final integration. However, the status of the decision circuit described, punched on to tape every 25 msec., would require a tape approaching 100 metres in length to define a typical chromatogram. For this reason, it is generally preferable to integrate each peak with a digital integrator connected to the chromatograph and to transfer only the peak areas, representing typically 20 values, to an off-line computer. To use the power of the off-line machine fully these values must be accompanied by corresponding elution times and intermediate baseline-level measurements.

Automatic baseline correction

To obtain meaningful results, each peak must be integrated with respect to the baseline of the chromatogram. The rate of deviation of the baseline from the true electrical zero depends on the type and size of the sample. A simple mechanical integrator operates as shown in Fig. 2(*a*), and provides no baseline correction. Most digital electronic integrators provide a correction as shown in Fig. 2(*b*) by integrating with respect to the last known baseline value before the peak. A further improvement can be made in more sophisticated instruments by taking into account the predicted shape of the baseline based on its previous variation (Fig. 2(*c*)). By use of an on-line

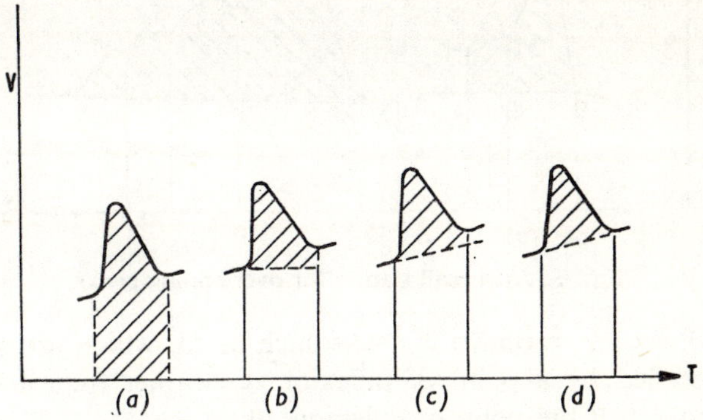

FIG. 2. Alternative methods of baseline correction.

computer (or an off-line machine provided with sufficient data) a retrospective correction can be made according to the baseline measurements before and after the peak (Fig. 2(*d*)). This form of correction is adequate for most analyses, but programmes can be written to allow for non-linear baseline deviation, when the drawing of a straight line under the peak is inadequate.

Area reallocation

The aim of the chromatographer should be to obtain an adequate component separation for all peaks, so that a baseline datum can be established for each one, by one of the methods described above. However, in analyses where two or more peaks overlap or small peaks elute on the tail of a large component, the digital computer provides a powerful method of deriving corrected areas by appropriate reallocation. Fig. 3(*a*) shows two overlapping peaks, of dissimilar size. A typical digital integrator might assign to these two peaks the areas ABCD and DCEFG, defined by the perpendiculars drawn from each trough. The digital computer programme, however, can effectively draw the line CF (a tangent to the chromatogram at F, drawn from the trough C) and add the area DCFG to that of the first peak. This procedure would normally be combined with baseline correction referred to the line AF rather than AG.

A similar technique is relevant for the small peaks D and F shown in Fig. 3(*b*), typical of a chromatogram with a large water vapour peak. The

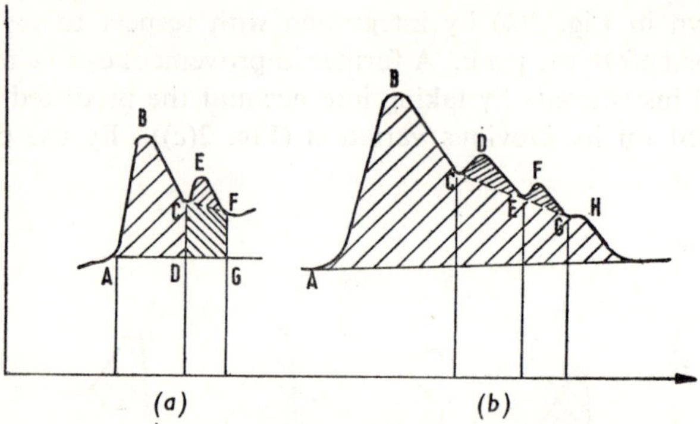

FIG. 3. Area reallocation for overlapping peaks.

method becomes less certain for peaks such as H, which are so small that they are preceded by a point of inflexion, G, rather than a trough. The computer can detect the point of inflexion, at

$$\frac{dV}{dt} = 0, \qquad \text{and} \qquad \frac{d^2V}{dt^2} = 0$$

However, in view of the amount of processing involved and the uncertainty of how to draw a division-line from the point G, such a programme may not always be justified. Not all overlapping peaks should be dealt with as described above—those of similar size are best separated by a perpendicular drawn from the trough and an area correction factor applied to each peak dependent on the size of the other. This is a field in which the human operator can often select the correct reallocation strategy far more reliably than the computer. Thus, for similar repeated analyses, it is best for the operator to "tag" each part of the chromatogram and allow the computer to do the work of area reallocation according to his decision.

Processing the area results

Most of the functions already described can be partly or wholly performed by an advanced digital integrator, but the computing power and flexibility of a stored programme digital computer are essential for subsequent stages. At the stage reached so far, the computer will have in its store a list of peak areas and corresponding elution times for each chromatogram to be analysed. (In the case of an off-line machine these data could be loaded from punched or magnetic tape for batch processing.) For each type of analysis the computer must be loaded with a set of data representing the chemical symbols or names, expected elution times and response factors for each component. It must also be instructed as to whether the peak areas are to be normalized, by dividing by the sum of all areas, or related instead to a reference component in the sample.

In an on-line system, reading and processing of data from several chromatographs proceeds simultaneously and a print-out is obtained immediately each analysis is complete. The actual form of report may vary considerably—in some cases only relative percentages of one or two components are required. For routine laboratory applications the reports may be produced on pre-printed forms and combined with other types of analysis.

Process control from the analysis

In some applications the analysis is repeated at frequent intervals, to monitor the result of a continuous reaction or experiment. The desired result can usually be expressed as a set of ratios between the peak heights or areas of the components of interest, for example

$$K_1 = \frac{A^1}{A^4}; \qquad K_2 = \frac{A^2}{A^4}$$

and so on.

The actual values of these ratios are computed on-line, from the corrected area results, and compared with preset desired values. In a simple system—for example control of a non-reacting mixture of gases—the ratio errors are used directly as control signals to adjust the inflow rates. In a more complex situation, each inflow or condition may interact with others and the computer must allow for these interactions, normally by reference to a simple mathematical mould of the process or reaction.

The control system itself is not a readily stable one, due to the time lags of the reaction and of the analysis, and conventional process control instruments may not be suitable. Good results have been obtained with a system that makes stepwise control adjustments to each input, of a definite computed size, immediately after each analysis, and waits for at least the plant time-constant period, before re-injecting the sample.

SUMMARY

Digital computers provide automatic on-line computation of percentage results from chromatographic analyses, and increase the power of the method by separating and correcting for unresolved peaks. In some applications the computer provides automatic control of a process or reaction monitored by a chromatograph.

DISCUSSION

Hill: What is the size of the computer you use for this work?

Marson: For a simple peak reallocation programme, without curve fitting, a K 8000-word system with a 12-bit or 16-bit machine (one of the small real-time machines available in the US) is adequate.

Lipsky: I am surprised that you need so many samples per second to characterize your curve. In high resolution mass spectrometry, for example, in which about 400 peaks are produced every ten seconds, our average sampling rate is about 20 to 30 points per curve. We have no difficulty in locating the sample at this rate.

Marson: The number of samples depends on whether the curve is assumed to be Gaussian. If it is, this number can be reduced to about five per second. But if the distribution is not Gaussian we have to increase the sampling rate, especially if we are looking for very small peaks eluting on a large curve.

Mitchell: Can Gaussian fits be done on-line? When working on a six-curve electrophoresis scan I found a programme was necessary which took 20 minutes of time on a very large computer.

Marson: Gaussian fits can be done on-line, but it is generally more economic to put the data on punched tape and process on a larger machine.

Mitchell: Is this system a practical proposition?

Marson: The process is working well in this country on one or two installations.

Scott: The sort of problems you described could surely have been dealt with equally well by a simple analogue system in conjunction with normal proportional and integral control. Probably three or four pertinent peaks only, from the chromatographic system, would provide sufficient data.

Marson: The system using an analogue peak-height store described at the end of my presentation is quite adequate in 95 per cent of applications. The digital method is most valuable when the analysis time has to be as short as possible and peak separation by programme must be used.

Scott: The real speed of operation is not related to the computer control system, whether this be analogue or digital, is it?

Marson: By using a post-analysis computation to separate overlapping peaks one can reduce analysis time by a useful margin.

Martin: How much do these systems cost?

Marson: The cost varies enormously. A simple analogue system working on peak heights can be and has been made for a few hundred pounds. A fairly small computer system, doing some peak reallocations, costs £10 000–£20 000. A machine capable of curve-fitting would generally be disc-based and cost about twice this amount.

Scott: Many of the calculations in gas chromatography can be carried out rapidly on a digital computer; retention indices and resolution are typical examples.

Hill: Our laboratory does many estimations of alcohol and blood anaesthetics and all the routine calculations are now put on to a computer, which takes a load off the technicians. I am thinking in terms of general laboratory chromatographs rather than of process control instruments. In chemical pathology laboratories in general hospitals in the UK an auto-analyser is now quite commonly coupled to a small on-line computer. The calculations needed for routine chemical pathology are trivial in themselves, but to measure out five thousand peaks on graph paper, as may be necessary in a busy hospital laboratory, is time-consuming and exhausting. Computers should be increasingly used to supplement manual operations. But what is ridiculous is that we still use a 1 μl. syringe to put our samples into the gas chromatograph. With an interpolation formula one can work with internal and external standards; this system works well for routine laboratory calculations.

Payne: One practical drawback of the auto-analyser/on-line computer system is the inability to extract from it a result needed urgently. One may have to wait several hours for the processing of an entire run to finish.

Hill: Provision can be made for putting in an extra sample; one loads the extra sample cup on at the front of the computer and it can be programmed to recognize this and give a rapid result.

Mitchell: There are two facets to the use of computers in clinical laboratories. On-line operation tends to be expensive and may have many built-in worries, particularly in clinical chemistry. A rather slower but often more practical approach is to feed a manual decision into the system at the optimum time. As you have said, Mr. Marson, it is often easier to pick out the peaks manually than by computer. If a curve-follower, on-line to the computer, is used, one can tear off the paper from the recorder of the gas chromatograph, auto-analyser or other instrument at suitable periods of time and quickly follow the curves or point the peaks, using the human brain to eliminate inequalities, delineate baseline drift and so on. The whole process is thereby made much cheaper and usually more efficient.

Marson: This is just how these machines should be used. One must compromise between the processing power of the computer and the decision-making capacity of the human brain.

Scott: In the Unilever Research Laboratories at Port Sunlight an infratonic system is used for routine analyses. The results are put out on a punched tape, this is fed into a computer and the results returned overnight. This is an effective method and relatively cheap to run.

Hill: Do you use a frequency modulation tape?

Scott: Yes. A tape that gives adequate performance for chromatography costs only about £200–300. We normally have about twenty chromatographs needing data handling and we have half-a-dozen tapes which we lend out to those who need them. The tapes are then taken to the infratonic integrator and the results either printed out or punched on to tape and fed to the computer.

Lowe: We have used a mechanical on-and-off signal for monitoring anaesthetic concentrations. This is a recorder with photocells located at fixed points on its height. The recording needle has an aluminium reflectant surface, and when it passes under the maximum concentration of anaesthetic it turns off the anaesthetic source; when the blood anaesthetic level drops below a pre-set known value, the needle turns the source on again. We have used this system to study a mixture of three anaesthetics given simultaneously to rats. We set the circuits with time delay switches, according to the retention times of the three anaesthetics, and were thus able to control inspired concentrations of all three anaesthetics simultaneously. Incidentally, the actual photocell arrangement for three anaesthetics really crowds the recorder slide on which six photocells are located. Microswitches could have been used in this system. Inspired halothane concentrations were regulated between 1·0 and 0·98 per cent.

Scott: This sounds like a good, reliable, simple technique.

Lowe: We have also controlled blood pressure within a fixed range during clinical anaesthesia. An arterial catheter was inserted into the patient's radial artery and a strain gauge used to monitor blood pressure between maximum and minimum limiting switches on a milliammeter.

Purnell: A galvanometer in the circuit would do the same job more cheaply.

Lowe: The system we used gives a permanent record as well as controlling concentrations. The galvanometer could not do this.

Hill: And the peaks give a simple check on the chromatograph column.

Payne: Do you actually keep the blood pressure within pre-set limits or do you just monitor it?

Lowe: We monitor the blood pressure, and then regulate anaesthetic concentrations with the maximum/minimum switches on the recorder. We bracket the switches round the required blood-pressure values and the recorder gives a signal for the injection of liquid anaesthetic. If the systolic blood pressure falls below, say, 90 mm. Hg the anaesthetic is turned off, and when the systolic pressure rises to 95 mm. Hg anaesthetic is injected again. This system will regulate and maintain the blood pressure within any chosen arbitrary limits.

Payne: This is fine provided the anaesthetic is the only parameter affecting the blood pressure. What happens if blood loss induces a fall in blood pressure?

Lowe: The system continues to regulate blood pressure whether this is being affected by anaesthetics or other factors such as bleeding.

GENERAL DISCUSSION

Sjövall: Dr. Brooks has already mentioned derivatives in connexion with forensic toxicology (pp. 146, 158). Derivatives yielding spectra showing the molecular ion can be made from most steroids. A molecular ion is not usually seen with steroids carrying several free hydroxyl groups at the high temperature required for gas chromatography.

Scott: I agree that derivatives are helpful in certain difficult analyses but we have to compromise between meticulous detail and speed. Of course it would be possible to trap all the components carefully and make derivatives but this might take more time than we can spare.

Brooks: One can make derivatives from various components of an entire mixture. The alcohol components in essential oils, for example, can all be converted to derivatives, leaving the hydrocarbons unchanged.

Scott: The essential oils are difficult to separate because of their complexity. Entirely new components may be reproduced by molecular rearrangement during the formation of derivatives. Some of these new components may be alcohols themselves. Thus one could easily end up with substances that bear no relation to the original components. The normal chromatographic separation of essential oils can easily produce substances that were never in the original mixture. This has nothing to do with identification techniques but shows that confusion can occur on the column alone.

Sjövall: And components of complex mixtures may react with each other, producing great confusion in the interpretation of the final gas chromatograms and mass spectra.

Scott: Chromatography of essential oils high-lights some problems of general interest. Some decomposition invariably occurs in most essential oils chromatographed at temperatures greater than 125° C. When we attempt to assess the odour of essential oils trapped from the column we find that a background odour is produced, due to this decomposition. And the decomposition odour can be so persistent as to make identification of the individual component of the column impossible. In liquid chromatography, however, decomposition and molecular rearrangement do not seem to occur to nearly the same extent as they do in gas chromatography. Essential oils also contain sulphides and mercaptans, which further complicates the analysis.

Brooks: Derivatives can stabilize some compounds and reduce their reactivity.

Scott: This is only true if one can make derivatives from each separate component, and if these individual components are available we do not often need derivatives anyhow.

Martin: I would like to continue our discussion on methods of chromatographic identifications. Almost all the commercially available chromatographs lack various elementary engineering refinements that would keep flow-rate and temperature constant. In consequence, the retention volume has acquired a bad name as a method of identification. Cramers (1967) has shown that retention volumes can be extremely accurately determined and, if this is done in a number of different solvents, precise identification is possible. One can thus guess the structure of many unknown compounds on inference, by comparison with suitable standards. Dr. J. J. Walraven and colleagues in Professor A. I. M. Keulemans' group at Eindhoven (Walraven, 1968; Walraven, Ladon and Keulemans, 1968) have used this technique but have so far only applied it to rather simple hydrocarbons. With two or three solvents one can easily identify every hydrocarbon up to the octanes and a large proportion of the nonanes. And one can usually recognize the presence of any extra non-hydrocarbon groups that have been introduced, thus amassing a lot of information about the molecule. If results from this technique were added to pyrolysis fragmentation patterns one could probably get as much information from these chromatographs as from a mass spectrometer. Research on the connexions between the molecular structure of the unknown substance and retention times of the fragmented materials would also be needed to develop this method. The large amount of basic research that still needs to be done here is not a good reason for not doing it. This is something which *should* be done. Various gadgets to perform chemistry on the individual peaks might be added to the chromatograph with advantage. (I am now departing from my previous view that one should keep instrumentation simple; I am, in general, in favour of gadgets!)

The method of molecular multiplication that Dr. Scott and I are currently studying could be arranged to produce figures for the analysis of carbon, hydrogen and possibly nitrogen for each peak as it comes out of the chromatograph. This would be another aid to identification. I predict that we can look forward to the gas chromatograph giving us a great deal of extra information in the fairly near future.

Carter: Would this system detect absolute amounts of carbon and hydrogen, or only ratios? And would it detect the presence of elements other than carbon and hydrogen?

Martin: It would detect absolute amounts. One would only be able to find individual elements such as oxygen, nitrogen, sulphur and phosphorus if the chromatograph was set up to look specifically for them. But the system will produce a ratio of any particular pair of elements for which it is set up.

Carter: If you set up the system to look for carbon and hydrogen, and oxygen were present, could you detect it?

Scott: No, but this would make no difference to the carbon to hydrogen ratio.

Carter: Would this system show the percentage of carbon in the test molecule?

Scott: No.

Martin: If a large sample of material is available, many chromatographic methods exist for determining the molecular weight. What we need is a reliable method for determining the molecular weights of very small quantities of material.

Lipsky: We could extrapolate the system you have just described to the analysis of complex mixtures such as cigarette smoke or coffee aroma, analysed on open-support columns. In this work adequate quantities of sample are available for the identification of peaks in certain areas, be it by retention times or retention volumes. If graphs of retention volumes were plotted against carbon numbers, and we knew the carbon and hydrogen content of each peak, the presence of a different molecule containing hetero atoms could be inferred. This information would be derived from the position of a particular peak in relation to the next peak containing one or two more carbon atoms. One could assume (after confirmation by other techniques) that two hetero atoms of one type or four hetero atoms of another type were present. This information would be particularly useful where present methods are excluded because of the limited size of the sample.

Carter: Is the column material itself a problem in chromatographic identification by retention times? Do the properties of the column change as a result of, say, slow oxidation of the stationary phase and, if so, does this affect relative retention times?

Martin: This inaccuracy can be detected by monitoring, at regular intervals, a set of test samples put down the column.

Carter: Has Keulemans' group been able to narrow down the choice of column material and recommend some as better than others for specific purposes?

Martin: Yes. Happily, people are now returning to my original view that it is desirable to choose substances as chemically different as possible for the stationary phases in order to look for different features in the test molecules.

Carter: Are some materials more stable, and hence more reliable, as the stationary phase than others?

Martin Dr. C. A. Cramers (unpublished) has worked with materials at quite low temperatures at which column stability is not a problem.

Purnell: I am enthusiastic about Professor Martin's recommendation to use a number of different column materials. We do not work with molecules as complex as the essential oils but normally we have to deal with molecules containing up to seven or eight carbon atoms. We use up to five columns for all our identifications and find that this covers most situations. For example we plot graphs of log retention volume against some empiric property such as boiling point or the number of characteristic atoms or groups in the molecule. The resultant lines for a series are not truly continuous but go from point to point since (for example) there is no hydrocarbon with carbon numbers between ethane and propane. Any unknown chemical has to be sited exactly on a standard reference point not just anywhere on a series line. If it is on the line but not exactly where a molecule should be it is not a member of that series.

Carter: I am surprised that you use as many as five column materials; what are they?

Purnell: We use squalane because it is essentially a boiling-point separator, oxydipropionitrile, polyethylene glycol 400, tricresyl phosphate and silicone oil. Gas chromatography is, in my view, a more powerful tool than any of the ancillary techniques, not excluding mass spectrometry which, although a beautiful technique, needs to be seen in proper perspective.

Scott: If, in general, you were considering a particular peak in a chromatogram of complex mixtures, how would you identify that peak when you rerun a mixture on a different stationary phase?

Purnell: This can be done by plotting log retention volumes for standard solutes with solvent "1" against those for solvent "2". Homologous series then give lines and, sometimes, curves. If the log retention volumes taken from the two chromatograms are marked off on the appropriate ordinate one can then draw perpendiculars to specific points. Pairs that intersect at a point for a standard solute are highly likely to be that substance. Note that they must intersect at a known solute point to give information. Any ambiguities remaining can be further resolved by the use of another solvent, and so on. Eventually every peak must be matched up on all chromatograms. When this matching is achieved it is surely more than coincidence and identifications ought to be sound.

Martin: How accurately are your points determined?

Purnell: We can easily measure retention volumes to an accuracy of about

1 per cent. The method reaches its limit at an accuracy of about 0·1 per cent because of fugacity variations in the carrier gas; these are minimally around 0·1–0·2 per cent for the usual type of system. At this stage the pressure compressibility corrections become so cumbersome that if greater accuracy is demanded the method stops being useful. The figure of 0·1 per cent accuracy for retention volumes does not apply to the regular chromatograph, of course, but only to an instrument specifically designed for this work.

Carter: What would happen if you started with a pure compound instead of a mixture? For example, could you take one peak from a chromatograph and pass it through different columns? This would avoid problems of separation and leave only problems concerning the measurement of time, and the gas compressibility correction would be smaller. Would this ease the fugacity problem? And could the column be made so short that pressure variations could be ignored?

Purnell: In principle one could work with columns of high porosity so that the pressure problem would be eliminated.

Martin: Or you could work very slowly.

Purnell: Yes, but even if we used a porous column for calibration we would not use it for analysis. The analytical column must generally be tightly packed with a high drop in pressure (about 101·3 kN/m^2. = 1 atm.) For the analyses of complex mixtures accuracy is finally limited by the efficiency of the analytical set-up.

Carter: I was visualizing one long column for analytical separation and other short columns to identify what had been separated. You could even have the same stationary phase in the analytical and one of the calibration columns, but under different conditions.

Purnell: This is what we normally do.

Blackmore: Our experience with the type of compounds met with in forensic toxicology, particularly on silicone gums, has been that retention times for samples of less than one microgramme vary with the amount of compound injected on the column. If a technique such as you have described, Professor Purnell, can measure retention volumes with an accuracy of 1 per cent, how can this degree of accuracy be equated with the quantity of material put on the column? Our work has been dogged by this difficulty; only a limited number of stationary phases do not suffer from this problem.

Gray: Do the *relative* retention volumes vary according to the amount of material put on the column? I suggest that different known concentrations of a "marker" substance put on the column at the same time as the test substances could serve as internal standards and solve this problem.

Blackmore: They do vary. Ten nanogrammes of a barbiturate, for example, will take about 5 per cent longer to emerge than ten microgrammes.

Martin: This would produce a distorted peak.

Blackmore: It is surprisingly undistorted.

Purnell: Conder (1969) has devised a fairly simple and quick method for calculating the true infinite dilution retention volume for any peak, irrespective of shape.

Blackmore: These sorts of approaches are all extremely relevant to problems in forensic toxicology.

Martin: The technique that Walraven (1968) has used to identify the different members of a group of similar substances is worth mentioning here. Consider the alkanes: each molecule is given a four-figure number. The first figure is the number of primary carbon atoms the molecule contains, the second the number of secondary atoms, the third the number of tertiary atoms and the fourth the number of quaternary atoms. If the log of the retention time, or the retention index, of the alkanes is plotted for two different stationary phases, all molecules with the same number fall on the same line and the lines form an imbricated series in the order of the numbers. It is easy to identify which line a substance belongs to and it is also usually easy to identify a particular isomer. Occasionally two lines are almost colinear but here the difficulty may be resolved by using a third solvent. This kind of scheme can be used for other classes of molecules as well as for the alkanes, although this makes the procedure more complicated. Nevertheless, with accurate data it is remarkable how easy identification often is.

A. Curry: It is important in forensic toxicology to be able to identify compounds that have similar pharmacological actions but are chemically unrelated. In one extract one may have to try to identify, on retention times from two, three or four columns, compounds as widely separated as amphetamine and nicotine, or diethyl propion and morphine. And these identifications have to be reliable enough for us to swear to their validity in court.

Martin: Would it help if you knew the ratios of carbon to hydrogen, carbon to nitrogen and carbon to sulphur?

A. Curry: It certainly would.

Lipsky: If the column eluent is split into two streams, one passing to a molecular multiplier and the other to a modified flame ionization detector containing an alkaline flame device, the latter could identify halogens and phosphorus in the unknown material.

Purnell: With all these techniques we have almost eliminated the need for a column!

Blackmore: In my presentation, Table III (p. 141) shows a mixture of decongestants, any of which may be extracted from a urine sample. This illustrates the tremendous chemical differences that can occur in a family of biologically similar compounds (antihistamines).

Martin: Can you obtain pure standards for all these substances?

Blackmore: Yes. We prepare our maps from samples of pure substances. We also find, when we use retention times, that unless we pick our columns very carefully we get different values for the retention volumes depending on the quantity present in the sample.

Martin: Obviously one should aim to work always at the ten microgramme level.

A. Curry: But we never know the quantity of toxic material that will be present in the sample.

Martin: You should run a second chromatograph to reduce to an appropriate quantity—ten microgrammes if possible.

Blackmore: These methods work reasonably well for some groups of compounds (barbiturates for example) at nanogramme concentrations. But when two or three antihistamines plus their metabolites, at unknown concentrations, are found in a single specimen of urine, the situation is more complex.

Payne: Before we finish our discussions, I would like to elicit some more information of two aspects of our subject: the use of gas chromatography in catecholamine analysis and the detection of steroids in routine clinical practice.

S. Curry: Gas chromatography is not yet an established tool for the study of catecholamines. Oxidation of catechol derivatives (to coloured materials) is probably responsible for the partial disappearance of catecholamines observed during extraction processes (Shore and Olin, 1958), and similar disappearance of these compounds on gas chromatography columns can be predicted. However, trifluoracetyl derivatives can be made and a method is known for the assay of 4-hydroxy, 3-methoxymandelic acid, a metabolite of adrenaline and noradrenaline (Wilk *et al.*, 1965). Students of catecholamines seem reluctant to use gas chromatography, although it would probably be highly applicable to their work.

Brooks: There are many methods of making catecholamine derivatives that show excellent gas chromatographic behaviour: some of these methods can be developed quantitatively. Other derivatives can be detected by the electron capture detector. But there is a great gap between the use of these methods for pure samples and for catecholamines from natural sources. Those scientists who are familiar with the isolation of natural catecholamines do not seem to have applied the gas-phase methods of analysis.

Payne: I get the impression that the methods currently available are still relatively inaccurate. Is that right?

Brooks: Yes.

S. Curry: Additional specificity and sensitivity are still needed in assays of catecholamines, especially in the study of these compounds in blood.

Brooks: But many of the available methods are quick and convenient, thus satisfying some of our clinical and biochemical needs.

Payne: Dr. Sjövall, will you comment on steroid detection by chromatography?

Sjövall: We are currently analysing steroid sulphates in plasma. These compounds are present in high concentration compared with the plasma concentrations of progesterone, testosterone or oestrogen and can be readily studied by gas chromatography.† Gas chromatographic methods for the latter compounds exist but could be greatly improved. There are several good methods for detecting the major urinary steroids. The book *Gas Phase Chromatography of Steroids* (Eik-Nes and Horning, 1968) contains much detailed and practical information about these techniques. One problem, as far as one can tell from many published chromatograms, is that at nanogramme sample sizes the columns used give too few plates to make the analyses convincingly specific. Purification of samples is another difficulty. Steroids present in low concentrations in tissues and body fluids are usually purified by column chromatography followed by thin layer chromatography. These steps have to be done repeatedly and spots have to be scraped off the plates; it would be simpler if more of the purification could be done by the gas chromatography column. Gas chromatography is often used as if it were a colour reaction: the sample is purified until one component remains and this is injected into the gas chromatograph. Although this is surely an unsuitable way of using gas chromatography such an approach may be needed because the concentrations of interfering compounds are so much higher than the concentration of the component being analysed.

S. Curry: Part of the problem may be that unwanted material in the chromatography support phase is washed from the chromatograms with the steroid. Thus a thin layer or paper chromatographic purification process can actually reduce the purity of the sample.

Sjövall: Yes. One must be very careful when using distilled solvents and washed chromatographic adsorbents. For the preliminary purification of steroids we are now trying liquid-gel chromatography on Sephadex derivatives which do not bleed and which can be used to obtain partition chromatographic systems that do not rely on the use of immiscible solvent systems.

† See also pp. 150–151.

The best derivatives we have found so far are long-chain ethers of Sephadex (Ellingboe, Nyström and Sjövall, 1968*a*, *b*). Depending on the solvent mixture used, these ethers form a stationary gel phase which is either more or less polar than the mobile phase. Nothing can bleed off these columns since the alkyl chains are covalently bound to the polysaccharide matrix. Fig. 1 shows a separation of steroids with this system, using heptane–chloroform, 8:2 (v/v), as the solvent. The peaks (from right to left) represent progesterone, 20β- and 20α-dihydroprogesterone, 11β- and 11α-hydroxyprogesterone, and corticosterone, eluted in order of increasing polarity (effluent volume is expressed as the number of total column volumes). Since the long-chain alkyl derivatives of Sephadex are hydrophobic they can be used in reversed

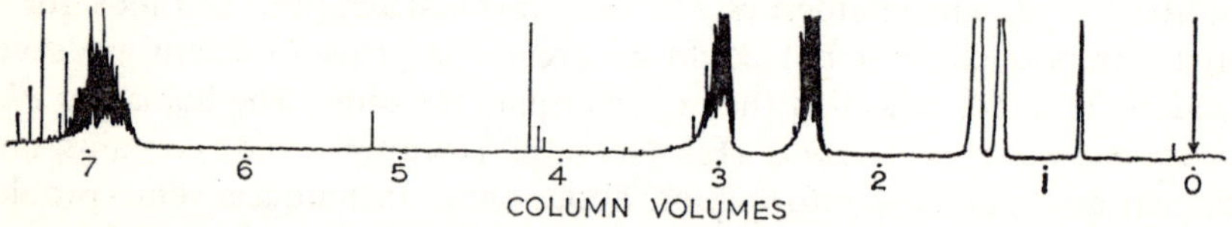

FIG. 1 (*Sjövall*). Separation of steroids using Sephadex derivatives (for explanation, see text).

phase chromatography, a technique to which I was first introduced through a paper by Professor Martin (Howard and Martin, 1950). Thus, cholesteryl esters differing by one methylene group can be separated on columns of the alkylated Sephadex gel in heptane, acetone and water in a ratio of 4 to 15 to 1 (Ellingboe, Nyström and Sjövall, 1968*a*, *b*).

We hope that liquid-gel chromatographic methods will be suitable for preliminary purification of steroids in urine samples to make them amenable to gas chromatographic and mass spectrometric analysis as groups rather than as individual components.

Mitchell: Gas chromatography in the analysis of steroids has not had the impact we expected. The gas chromatograph is rarely used for routine purposes in the major steroid laboratories of the world, although it is used for research. The reason for this is that steroids can be measured colorimetrically in fairly impure mixtures because of the high specificity of the colour reactions that can be elicited. This cannot be done using the non-specific detectors available in gas chromatography. Oestrogen assays are typical in that liquid chromatographic methods are available for oestrogen assay in urine and in blood from non-pregnant women (Loraine and Bell, 1966); but although workers have tried to use gas chromatographic techniques for these assays the older methods have not yet been bettered. When more specific detectors become available gas chromatography could have a considerable impact on the detection of the sex hormones.

Lipsky: The main problem is not the detection of individual steroids but putting the work on a reliable routine basis in clinical laboratories. Purification is very difficult when one is looking for fractions of one microgramme per litre in a total sample of one or two litres of urine. Quite adequate gas chromatographic methods for the analysis of steroid derivatives *are* now available. The problem is in handling the sample and changing it into a form in which it can be introduced into the chromatograph.

Martin: Groenendijk and co-workers (1969) in Keulemans' laboratory are using capillary columns; their work shows some real advances in converting gas chromatography of steroids into a routine process. These workers are now getting good resolution with nanogramme quantities of steroid derivatives.

REFERENCES

Conder, J. R. (1969). *J. Chromat.*, **39**, 273–281.

Cramers, C. A. (1967). Ph.D. Thesis. Technological University of Eindhoven.

Eik-Nes, K. B., and Horning, E. C. (eds) (1968). *Gas Phase Chromatography of Steroids.* Berlin: Springer Verlag.

Ellingboe, J., Nyström, E., and Sjövall, J. (1968*a*). *Biochim. biophys. Acta*, **152,** 803–805.

Ellingboe, J., Nyström, E., and Sjövall, J. (1968*b*). *Fedn Proc. Fedn Am. Socs exp. Biol.*, **27,** 648.

Groenendijk, H., and van Kemenade, A. W. C. (1969). *Chromatographia*, **3,** 107–110.

Howard, G. A., and Martin, A. J. P. (1950). *Biochem. J.*, **46,** 532–538.

Loraine, J. A., and Bell, E. T. (eds.) (1966). *Hormone Assays and their Clinical Application.* London: Livingstone.

Shore, P. A., and Olin, J. S. (1958). *J. Pharmac. exp. Ther.*, **122,** 295–300.

Walraven, J. J. (1968). Ph.D. Thesis. Technological University of Eindhoven.

Walraven, J. J., Ladon, A. W., and Keulemans, A. I. M. (1968). *Chromatographia*, **1,** 195–198.

Wilk, S., Gittow, S. E., Mendlowitz, M., Franklin, M. J., and Carr, H. E. (1965). *Analyt. Biochem.*, **13,** 544–551.

CHAIRMAN'S CLOSING REMARKS

J. P. PAYNE

I HAVE no intention of summarizing all that has been said during the past two days but there are one or two points worth making. The clinicians among us must have derived some satisfaction from the realization that many of the problems to be overcome, for example in the collection and handling of blood samples, are technical rather than clinical, and that they (the clinicians) are not entirely to blame for failing to make the best use of the techniques of gas chromatography. The chemists, engineers and physicists for their part may well have seen the problems of the clinicians in a new perspective and, as Professor Purnell pointed out, perhaps these problems should be looked at from the point of view of industrial instrument processing. This is a new concept for many of us who work in the clinical field but it may be a most useful approach to the problems of monitoring and sample handling in clinical medicine.

Perhaps the highlight of this meeting was the account by Professor Martin of the early development of chromatography, a field which must now be recognized as one of the most powerful and versatile available to the biological scientist.

During the past two days gas chromatographic methods for the detection and quantitative evaluation of physiological blood gases, anaesthetics, catecholamines, steroids and poisons have been discussed, and the list is far from complete. What has become increasingly apparent is that, as well as versatility, gas chromatography offers a degree of reproducibility and accuracy in biological measurement unsurpassed by any other available technique. When considerations of cost are also included the place of the technique in biology and medicine seems assured.

On your behalf I would like to thank the Trustees of the Ciba Foundation and Dr. Wolstenholme and his staff not only for making this symposium possible but also for making the whole experience so enjoyable. Finally, as chairman, I owe you all a personal debt of gratitude for making my task so easy. I have not had to separate warring protagonists nor have I needed to interrupt the proceedings except to make my own voice heard, and for this I am truly grateful.

INDEX OF AUTHORS*

Numbers in bold type indicate a contribution in the form of a paper; numbers in plain type refer to contributions to the discussions.

* Author and Subject Indexes compiled by Mr. William Hill.

INDEX OF SUBJECTS

Printed in Great Britain by Spottiswoode, Ballantyne & Co. Ltd., London and Colchester